The Science of Football

Physiology, Biomechanics, Tactics, Performance Analysis, and Injury Prevention in the Modern Game

Rajeed S. Kumar

The Science of Football

Physiology, Biomechanics, Tactics, Performance Analysis, and Injury Prevention in the Modern Game

ELIVA PRESS

ISBN: 978-99993-4-918-5

Cover Design: Eliva Press
Cover Image: www.ingimage.com
Email: info@elivapress.com
Website: www.elivapress.com

This Eliva Press imprint is published by the registered company Eliva Press Global Ltd. part of Eliva Press S.R.L. Publishing Group

The registered company address is: Pope Hennessy Street Level 2, Hennessy Tower Port Louis, Mauritius
Eliva Press S.R.L. Publishing Group legal address is: Bd. Cuza-Voda 1/4 of. 21, Chisinau, Moldova, Europe

The Science of Football

Physiology, Biomechanics, Tactics, Performance Analysis, and Injury Prevention in the Modern Game

Eliva Press Academic Series

A scientific and practical introduction to football performance for students, coaches, researchers, analysts, and informed readers.

2026 Edition

Copyright and Use

Book Description

The Science of Football examines the world's most popular sport as a demanding performance system. It explains how players run, accelerate, kick, scan, decide, recover, and cooperate inside the tactical structure of a team.

The book connects physiology, biomechanics, tactical theory, match analysis, psychology, sports medicine, youth development, women's football, and emerging technology. It is written for readers who want an academic but practical understanding of modern football.

Rather than reducing the game to statistics or romantic language, this book shows how scientific knowledge can deepen appreciation of football. The goal is to make performance visible: why high-speed running matters, why tactical spacing changes physical demand, why injury prevention requires neuromuscular preparation, and why data must be interpreted with context.

Contents

Chapter 1

Football as a Scientific System

This chapter introduces football as an object of scientific investigation. The goal is to show why the sport demands a systems perspective, where physiology, skill, tactics, psychology, and technology are treated as interacting dimensions of performance.

1.1 The Game as a Complex Performance Environment

Performance analysis converts match events into interpretable information. Event data records actions such as passes, shots, tackles, and carries, while tracking data captures player and ball positions over time. Together they allow analysts to study not only what happened, but also where, when, and under which spatial conditions it happened.

Expected goals models estimate the probability that a shot becomes a goal based on historical patterns and shot characteristics. The metric is useful because it separates chance quality from finishing outcome, but it must be interpreted carefully. Models differ in their inputs, and a single match can be strongly shaped by variance.

Data becomes valuable only when it supports better questions. A dashboard full of numbers can distract a coaching staff if the metrics are disconnected from the team's model of play. Good analysis should clarify decisions, not replace football judgement.

The strongest programmes are iterative. They observe performance, identify problems, design interventions, test them in training, and adjust them according to evidence and player response. This cycle resembles scientific method, but it must operate at the speed of a competitive season.

Transitions are decisive because teams are temporarily disorganized when possession changes. The first few seconds after losing or winning the ball can determine whether a team creates a chance, prevents a counterattack, or loses control of the match.

Consider an academy player who dominates because of early maturation. The scouting report should distinguish current physical advantage from adaptable skill, perception, learning behaviour, and resilience. Long-term development depends on identifying qualities that will still matter when physical differences narrow.

The best evidence-informed practice combines published research, local data, practitioner expertise, and player feedback. None of these sources is sufficient on its own.

Decision-making in football is time pressured and information rich. Players must scan before receiving, anticipate movement, recognize patterns, and decide while opponents attempt to hide cues or create deception. Cognitive speed is therefore inseparable from technical speed.

In practical terms, the game as a complex performance environment should be understood as part of the larger performance ecology of football. The scientific task is to describe the relevant mechanisms, while the coaching task is to transform that knowledge into training, selection, communication, and match decisions.

1.2 Scientific Method in Football Research

Tactics organize individual actions into collective behaviour. Formations provide reference points, but they do not describe the full reality of play. A team may defend in one structure, build attacks in another, press in a third, and alter spacing according to the opponent's behaviour.

Pressing is not simply running at the ball. Effective pressing requires triggers, cover shadows, distances between lines, timing from supporting players, and a clear plan for what happens if the press is broken. Poorly organized pressing can create large spaces behind the first line of pressure.

Transitions are decisive because teams are temporarily disorganized when possession changes. The first few seconds after losing or winning the ball can determine whether a team creates a chance, prevents a counterattack, or loses control of the match.

Context should protect practitioners from false certainty. A metric that is useful for one team model may be misleading for another. A player who appears inefficient in one tactical structure may become highly valuable in a different role. Scientific interpretation must therefore remain conditional and specific.

Decision-making in football is time pressured and information rich. Players must scan before receiving, anticipate movement, recognize patterns, and decide while opponents attempt to hide cues or create deception. Cognitive speed is therefore inseparable from technical speed.

Consider a wide player who repeatedly makes high-speed runs behind the defensive line. A physical report may describe sprint distance, but tactical analysis must ask whether the runs were timed well, whether teammates had passing access, whether the runs stretched the opponent, and whether the player could repeat them after fatigue. The scientific value appears when physical output is connected to tactical effect.

Many football studies are observational. They describe relationships but do not always prove causality. A variable may correlate with success because good teams create the conditions for it, not because the variable alone produces success.

The scientific study of football must balance reduction and integration. Reduction allows the researcher to isolate sprint speed, joint angle, oxygen uptake, or passing frequency. Integration asks how these variables function inside a match. Both approaches are necessary; the danger lies in treating an isolated variable as if it explains performance by itself.

In practical terms, scientific method in football research should be understood as part of the larger performance ecology of football. The scientific task is to describe the relevant mechanisms, while the coaching task is to transform that knowledge into training, selection, communication, and match decisions.

1.3 The Player as a Biological System

Decision-making in football is time pressured and information rich. Players must scan before receiving, anticipate movement, recognize patterns, and decide while opponents attempt to hide cues or create deception. Cognitive speed is therefore inseparable from technical speed.

Confidence affects performance because it influences attention, risk tolerance, and emotional control. Under pressure, players may narrow their focus, rush decisions, or avoid demanding actions. Psychological preparation helps players maintain useful attention and commit to appropriate decisions in unstable moments.

Team cohesion is not a vague emotional quality. It appears in communication, pressing commitment, recovery runs, role acceptance, and the willingness to act for collective benefit. A team with strong cohesion can execute demanding tactical plans more consistently.

For applied work, the central question is not whether the variable can be measured, but whether it changes decisions. A coach, analyst, or sports scientist should ask how the

information affects training design, match preparation, player development, or medical risk management. Measurement without decision relevance adds administrative burden rather than performance value.

A match is produced by interacting constraints: rules, pitch dimensions, weather, fatigue, the referee's interpretation, coaching strategy, crowd pressure, and the changing distribution of space. These constraints do not operate separately. They continuously reshape the options available to each player and determine whether a technical action is easy, risky, delayed, or impossible.

Consider a central midfielder who completes many short passes. A raw passing percentage may look excellent, but the analyst must examine pressure, pass direction, receiving body shape, and whether the passes progressed play. Football data becomes meaningful when it differentiates safe circulation from actions that alter the opponent's defensive structure.

The best evidence-informed practice combines published research, local data, practitioner expertise, and player feedback. None of these sources is sufficient on its own.

Expected goals models estimate the probability that a shot becomes a goal based on historical patterns and shot characteristics. The metric is useful because it separates chance quality from finishing outcome, but it must be interpreted carefully. Models differ in their inputs, and a single match can be strongly shaped by variance.

In practical terms, the player as a biological system should be understood as part of the larger performance ecology of football. The scientific task is to describe the relevant mechanisms, while the coaching task is to transform that knowledge into training, selection, communication, and match decisions.

1.4 The Team as a Dynamic System

Football is a complex adaptive sport rather than a closed laboratory task. The same pass, sprint, tackle, or shot can have different value depending on the score, the opponent's shape, the phase of play, the time in the match, and the individual qualities of the players involved. Scientific analysis is therefore most useful when it respects the ecology of the game and connects measurement to tactical meaning.

A match is produced by interacting constraints: rules, pitch dimensions, weather, fatigue, the referee's interpretation, coaching strategy, crowd pressure, and the changing distribution of space. These constraints do not operate separately. They continuously reshape the options available to each player and determine whether a technical action is easy, risky, delayed, or impossible.

The scientific study of football must balance reduction and integration. Reduction allows the researcher to isolate sprint speed, joint angle, oxygen uptake, or passing frequency. Integration asks how these variables function inside a match. Both approaches are necessary; the danger lies in treating an isolated variable as if it explains performance by itself.

A practical football environment also requires shared language. Coaches may describe a situation tactically, medical staff may describe it through load or risk, analysts may describe it through spatial data, and players may describe it through feel. The interdisciplinary task is to translate these perspectives into a coherent plan.

Data becomes valuable only when it supports better questions. A dashboard full of numbers can distract a coaching staff if the metrics are disconnected from the team's model of play. Good analysis should clarify decisions, not replace football judgement.

Consider a defender returning from a hamstring injury. Laboratory strength tests are useful, but they do not fully represent the demands of chasing an attacker, braking sharply, turning under pressure, and clearing the ball while fatigued. Return-to-play should therefore progress from controlled tests to football-specific exposure.

Many football studies are observational. They describe relationships but do not always prove causality. A variable may correlate with success because good teams create the conditions for it, not because the variable alone produces success.

Tactics organize individual actions into collective behaviour. Formations provide reference points, but they do not describe the full reality of play. A team may defend in one structure, build attacks in another, press in a third, and alter spacing according to the opponent's behaviour.

In practical terms, the team as a dynamic system should be understood as part of the larger performance ecology of football. The scientific task is to describe the relevant mechanisms, while the coaching task is to transform that knowledge into training, selection, communication, and match decisions.

1.5 Evidence, Context, and Coaching Judgement

Performance analysis converts match events into interpretable information. Event data records actions such as passes, shots, tackles, and carries, while tracking data captures player and ball positions over time. Together they allow analysts to study not only what happened, but also where, when, and under which spatial conditions it happened.

Expected goals models estimate the probability that a shot becomes a goal based on historical patterns and shot characteristics. The metric is useful because it separates chance quality from finishing outcome, but it must be interpreted carefully. Models differ in their inputs, and a single match can be strongly shaped by variance.

Data becomes valuable only when it supports better questions. A dashboard full of numbers can distract a coaching staff if the metrics are disconnected from the team's model of play. Good analysis should clarify decisions, not replace football judgement.

The strongest programmes are iterative. They observe performance, identify problems, design interventions, test them in training, and adjust them according to evidence and player response. This cycle resembles scientific method, but it must operate at the speed of a competitive season.

Tactics organize individual actions into collective behaviour. Formations provide reference points, but they do not describe the full reality of play. A team may defend in one structure, build attacks in another, press in a third, and alter spacing according to the opponent's behaviour.

Consider an academy player who dominates because of early maturation. The scouting report should distinguish current physical advantage from adaptable skill, perception, learning behaviour, and resilience. Long-term development depends on identifying qualities that will still matter when physical differences narrow.

The best evidence-informed practice combines published research, local data, practitioner expertise, and player feedback. None of these sources is sufficient on its own.

Team cohesion is not a vague emotional quality. It appears in communication, pressing commitment, recovery runs, role acceptance, and the willingness to act for collective benefit. A team with strong cohesion can execute demanding tactical plans more consistently.

In practical terms, evidence, context, and coaching judgement should be understood as part of the larger performance ecology of football. The scientific task is to describe the relevant mechanisms, while the coaching task is to transform that knowledge into training, selection, communication, and match decisions.

1.6 Practical Applications for Modern Football

Tactics organize individual actions into collective behaviour. Formations provide reference points, but they do not describe the full reality of play. A team may defend in one structure, build attacks in another, press in a third, and alter spacing according to the opponent's behaviour.

Pressing is not simply running at the ball. Effective pressing requires triggers, cover shadows, distances between lines, timing from supporting players, and a clear plan for what happens if the press is broken. Poorly organized pressing can create large spaces behind the first line of pressure.

Transitions are decisive because teams are temporarily disorganized when possession changes. The first few seconds after losing or winning the ball can determine whether a team creates a chance, prevents a counterattack, or loses control of the match.

Context should protect practitioners from false certainty. A metric that is useful for one team model may be misleading for another. A player who appears inefficient in one tactical structure may become highly valuable in a different role. Scientific interpretation must therefore remain conditional and specific.

Confidence affects performance because it influences attention, risk tolerance, and emotional control. Under pressure, players may narrow their focus, rush decisions, or avoid demanding actions. Psychological preparation helps players maintain useful attention and commit to appropriate decisions in unstable moments.

Consider a wide player who repeatedly makes high-speed runs behind the defensive line. A physical report may describe sprint distance, but tactical analysis must ask whether the runs were timed well, whether teammates had passing access, whether the runs stretched the opponent, and whether the player could repeat them after fatigue. The scientific value appears when physical output is connected to tactical effect.

Many football studies are observational. They describe relationships but do not always prove causality. A variable may correlate with success because good teams create the conditions for it, not because the variable alone produces success.

A match is produced by interacting constraints: rules, pitch dimensions, weather, fatigue, the referee's interpretation, coaching strategy, crowd pressure, and the changing distribution of space. These constraints do not operate separately. They continuously reshape the options available to each player and determine whether a technical action is easy, risky, delayed, or impossible.

In practical terms, practical applications for modern football should be understood as part of the larger performance ecology of football. The scientific task is to describe the relevant mechanisms, while the coaching task is to transform that knowledge into training, selection, communication, and match decisions.

Chapter 2

Physiological Demands of Match Play

This chapter examines the physical requirements of match play and training. Football demands endurance, repeated high-intensity capacity, speed, strength, and recovery, but these qualities must be developed in ways that serve the tactical model of the team.

2.1 Intermittent Exercise and Energy Systems

Football injuries emerge from the interaction between tissue capacity, movement demands, fatigue, contact, history, and training load. No single variable explains risk. Prevention therefore requires a system that combines strength, neuromuscular control, appropriate workload progression, recovery, and clear communication between coaches and medical staff.

Hamstring injuries are common because football repeatedly exposes players to sprinting, high-speed running, and rapid changes in muscle length. Prevention strategies usually combine sprint exposure, eccentric strength, progressive loading, and careful monitoring after congested match periods.

Return-to-play decisions should not be based only on the absence of pain. A player may be pain-free yet not ready for the speed, contact, fatigue, and decision-making demands of competition. The final stage of rehabilitation must recreate football-specific stress in a controlled progression.

Context should protect practitioners from false certainty. A metric that is useful for one team model may be misleading for another. A player who appears inefficient in one tactical structure may become highly valuable in a different role. Scientific interpretation must therefore remain conditional and specific.

Performance analysis converts match events into interpretable information. Event data records actions such as passes, shots, tackles, and carries, while tracking data captures player and ball positions over time. Together they allow analysts to study not only what happened, but also where, when, and under which spatial conditions it happened.

Consider a central midfielder who completes many short passes. A raw passing percentage may look excellent, but the analyst must examine pressure, pass direction, receiving body shape, and whether the passes progressed play. Football data becomes meaningful when it differentiates safe circulation from actions that alter the opponent's defensive structure.

Research evidence should be read with attention to sample size, playing level, age group, sex, competitive context, and measurement technology. Findings from elite male professionals do not always transfer directly to youth football, amateur environments, or women's football.

Pressing is not simply running at the ball. Effective pressing requires triggers, cover shadows, distances between lines, timing from supporting players, and a clear plan for what happens if the press is broken. Poorly organized pressing can create large spaces behind the first line of pressure.

In practical terms, intermittent exercise and energy systems should be understood as part of the larger performance ecology of football. The scientific task is to describe the relevant mechanisms, while the coaching task is to transform that knowledge into training, selection, communication, and match decisions.

2.2 Aerobic Capacity and Recovery Between Efforts

Performance analysis converts match events into interpretable information. Event data records actions such as passes, shots, tackles, and carries, while tracking data captures player and ball positions over time. Together they allow analysts to study not only what happened, but also where, when, and under which spatial conditions it happened.

Expected goals models estimate the probability that a shot becomes a goal based on historical patterns and shot characteristics. The metric is useful because it separates chance quality from finishing outcome, but it must be interpreted carefully. Models differ in their inputs, and a single match can be strongly shaped by variance.

Data becomes valuable only when it supports better questions. A dashboard full of numbers can distract a coaching staff if the metrics are disconnected from the team's model of play. Good analysis should clarify decisions, not replace football judgement.

For applied work, the central question is not whether the variable can be measured, but whether it changes decisions. A coach, analyst, or sports scientist should ask how the information affects training design, match preparation, player development, or medical risk management. Measurement without decision relevance adds administrative burden rather than performance value.

Pressing is not simply running at the ball. Effective pressing requires triggers, cover shadows, distances between lines, timing from supporting players, and a clear plan for what happens if the press is broken. Poorly organized pressing can create large spaces behind the first line of pressure.

Consider a defender returning from a hamstring injury. Laboratory strength tests are useful, but they do not fully represent the demands of chasing an attacker, braking sharply, turning under pressure, and clearing the ball while fatigued. Return-to-play should therefore progress from controlled tests to football-specific exposure.

Thresholds for speed zones, workload alerts, and fatigue markers vary across studies and commercial systems. Practitioners should understand how a threshold is defined before comparing numbers across teams, leagues, or seasons.

Football performance depends on repeated high-intensity actions embedded within long periods of lower-intensity movement. Players walk, jog, accelerate, decelerate, sprint, jump, collide, and recover, often within short time windows. The energy cost of the game is therefore not represented adequately by total distance alone; the pattern and timing of intense actions are equally important.

In practical terms, aerobic capacity and recovery between efforts should be understood as part of the larger performance ecology of football. The scientific task is to describe the relevant mechanisms, while the coaching task is to transform that knowledge into training, selection, communication, and match decisions.

2.3 Anaerobic Power, Sprinting, and Repeated Actions

Tactics organize individual actions into collective behaviour. Formations provide reference points, but they do not describe the full reality of play. A team may defend in one structure, build attacks in another, press in a third, and alter spacing according to the opponent's behaviour.

Pressing is not simply running at the ball. Effective pressing requires triggers, cover shadows, distances between lines, timing from supporting players, and a clear plan for what happens if the press is broken. Poorly organized pressing can create large spaces behind the first line of pressure.

Transitions are decisive because teams are temporarily disorganized when possession changes. The first few seconds after losing or winning the ball can determine whether a team creates a chance, prevents a counterattack, or loses control of the match.

A practical football environment also requires shared language. Coaches may describe a situation tactically, medical staff may describe it through load or risk, analysts may describe it through spatial data, and players may describe it through feel. The interdisciplinary task is to translate these perspectives into a coherent plan.

Physiological preparation must respect positional demands. Full-backs and wide midfielders may require repeated high-speed running along corridors of the pitch, centre-backs may face intense accelerations and aerial duels, and central midfielders often accumulate large volumes of medium-intensity movement while constantly adjusting position.

Consider an academy player who dominates because of early maturation. The scouting report should distinguish current physical advantage from adaptable skill, perception, learning behaviour, and resilience. Long-term development depends on identifying qualities that will still matter when physical differences narrow.

Research evidence should be read with attention to sample size, playing level, age group, sex, competitive context, and measurement technology. Findings from elite male professionals do not always transfer directly to youth football, amateur environments, or women's football.

Return-to-play decisions should not be based only on the absence of pain. A player may be pain-free yet not ready for the speed, contact, fatigue, and decision-making demands of competition. The final stage of rehabilitation must recreate football-specific stress in a controlled progression.

In practical terms, anaerobic power, sprinting, and repeated actions should be understood as part of the larger performance ecology of football. The scientific task is to describe the relevant mechanisms, while the coaching task is to transform that knowledge into training, selection, communication, and match decisions.

2.4 High-Speed Running and Positional Demands

Football performance depends on repeated high-intensity actions embedded within long periods of lower-intensity movement. Players walk, jog, accelerate, decelerate, sprint, jump, collide, and recover, often within short time windows. The energy cost of the game is therefore not represented adequately by total distance alone; the pattern and timing of intense actions are equally important.

Aerobic fitness supports football because it helps players recover between explosive efforts, maintain technical quality during fatigue, and tolerate demanding training blocks. However, football is not a simple endurance sport. The decisive moments frequently arise from brief accelerations, rapid changes of direction, high-speed runs into space, and repeated pressing actions.

Physiological preparation must respect positional demands. Full-backs and wide midfielders may require repeated high-speed running along corridors of the pitch, centre-backs may face intense accelerations and aerial duels, and central midfielders often accumulate large volumes of medium-intensity movement while constantly adjusting position.

The strongest programmes are iterative. They observe performance, identify problems, design interventions, test them in training, and adjust them according to evidence and player response. This cycle resembles scientific method, but it must operate at the speed of a competitive season.

Football injuries emerge from the interaction between tissue capacity, movement demands, fatigue, contact, history, and training load. No single variable explains risk. Prevention therefore requires a system that combines strength, neuromuscular control, appropriate workload progression, recovery, and clear communication between coaches and medical staff.

Consider a wide player who repeatedly makes high-speed runs behind the defensive line. A physical report may describe sprint distance, but tactical analysis must ask whether the runs were timed well, whether teammates had passing access, whether the runs stretched the opponent, and whether the player could repeat them after fatigue. The scientific value appears when physical output is connected to tactical effect.

Thresholds for speed zones, workload alerts, and fatigue markers vary across studies and commercial systems. Practitioners should understand how a threshold is defined before comparing numbers across teams, leagues, or seasons.

Expected goals models estimate the probability that a shot becomes a goal based on historical patterns and shot characteristics. The metric is useful because it separates chance quality from finishing outcome, but it must be interpreted carefully. Models differ in their inputs, and a single match can be strongly shaped by variance.

In practical terms, high-speed running and positional demands should be understood as part of the larger performance ecology of football. The scientific task is to describe the relevant mechanisms, while the coaching task is to transform that knowledge into training, selection, communication, and match decisions.

2.5 Fatigue Across Halves and Match Congestion

Football injuries emerge from the interaction between tissue capacity, movement demands, fatigue, contact, history, and training load. No single variable explains risk. Prevention therefore requires a system that combines strength, neuromuscular control, appropriate workload progression, recovery, and clear communication between coaches and medical staff.

Hamstring injuries are common because football repeatedly exposes players to sprinting, high-speed running, and rapid changes in muscle length. Prevention strategies usually combine sprint exposure, eccentric strength, progressive loading, and careful monitoring after congested match periods.

Return-to-play decisions should not be based only on the absence of pain. A player may be pain-free yet not ready for the speed, contact, fatigue, and decision-making demands of competition. The final stage of rehabilitation must recreate football-specific stress in a controlled progression.

Context should protect practitioners from false certainty. A metric that is useful for one team model may be misleading for another. A player who appears inefficient in one tactical structure may become highly valuable in a different role. Scientific interpretation must therefore remain conditional and specific.

Expected goals models estimate the probability that a shot becomes a goal based on historical patterns and shot characteristics. The metric is useful because it separates chance quality from finishing outcome, but it must be interpreted carefully. Models differ in their inputs, and a single match can be strongly shaped by variance.

Consider a central midfielder who completes many short passes. A raw passing percentage may look excellent, but the analyst must examine pressure, pass direction, receiving body shape, and whether the passes progressed play. Football data becomes meaningful when it differentiates safe circulation from actions that alter the opponent's defensive structure.

Research evidence should be read with attention to sample size, playing level, age group, sex, competitive context, and measurement technology. Findings from elite male professionals do not always transfer directly to youth football, amateur environments, or women's football.

Tactics organize individual actions into collective behaviour. Formations provide reference points, but they do not describe the full reality of play. A team may defend in one structure, build attacks in another, press in a third, and alter spacing according to the opponent's behaviour.

In practical terms, fatigue across halves and match congestion should be understood as part of the larger performance ecology of football. The scientific task is to describe the relevant mechanisms, while the coaching task is to transform that knowledge into training, selection, communication, and match decisions.

2.6 Monitoring Load in Training and Competition

Performance analysis converts match events into interpretable information. Event data records actions such as passes, shots, tackles, and carries, while tracking data captures player and ball positions over time. Together they allow analysts to study not only what happened, but also where, when, and under which spatial conditions it happened.

Expected goals models estimate the probability that a shot becomes a goal based on historical patterns and shot characteristics. The metric is useful because it separates chance quality from finishing outcome, but it must be interpreted carefully. Models differ in their inputs, and a single match can be strongly shaped by variance.

Data becomes valuable only when it supports better questions. A dashboard full of numbers can distract a coaching staff if the metrics are disconnected from the team's model of play. Good analysis should clarify decisions, not replace football judgement.

For applied work, the central question is not whether the variable can be measured, but whether it changes decisions. A coach, analyst, or sports scientist should ask how the information affects training design, match preparation, player development, or medical risk management. Measurement without decision relevance adds administrative burden rather than performance value.

Transitions are decisive because teams are temporarily disorganized when possession changes. The first few seconds after losing or winning the ball can determine whether a team creates a chance, prevents a counterattack, or loses control of the match.

Consider a defender returning from a hamstring injury. Laboratory strength tests are useful, but they do not fully represent the demands of chasing an attacker, braking sharply, turning under pressure, and clearing the ball while fatigued. Return-to-play should therefore progress from controlled tests to football-specific exposure.

Thresholds for speed zones, workload alerts, and fatigue markers vary across studies and commercial systems. Practitioners should understand how a threshold is defined before comparing numbers across teams, leagues, or seasons.

Physiological preparation must respect positional demands. Full-backs and wide midfielders may require repeated high-speed running along corridors of the pitch, centre-backs may face intense accelerations and aerial duels, and central midfielders often accumulate large volumes of medium-intensity movement while constantly adjusting position.

In practical terms, monitoring load in training and competition should be understood as part of the larger performance ecology of football. The scientific task is to describe the relevant mechanisms, while the coaching task is to transform that knowledge into training, selection, communication, and match decisions.

Chapter 3

Biomechanics of Football Movement

This chapter explains how movement mechanics shape football performance. Biomechanics provides a language for understanding running, kicking, braking, jumping, landing, and changing direction, all of which influence both performance and injury risk.

3.1 Running Mechanics in Football Contexts

Football injuries emerge from the interaction between tissue capacity, movement demands, fatigue, contact, history, and training load. No single variable explains risk. Prevention therefore requires a system that combines strength, neuromuscular control, appropriate workload progression, recovery, and clear communication between coaches and medical staff.

Hamstring injuries are common because football repeatedly exposes players to sprinting, high-speed running, and rapid changes in muscle length. Prevention strategies usually combine sprint exposure, eccentric strength, progressive loading, and careful monitoring after congested match periods.

Return-to-play decisions should not be based only on the absence of pain. A player may be pain-free yet not ready for the speed, contact, fatigue, and decision-making demands of competition. The final stage of rehabilitation must recreate football-specific stress in a controlled progression.

For applied work, the central question is not whether the variable can be measured, but whether it changes decisions. A coach, analyst, or sports scientist should ask how the information affects training design, match preparation, player development, or medical risk management. Measurement without decision relevance adds administrative burden rather than performance value.

Aerobic fitness supports football because it helps players recover between explosive efforts, maintain technical quality during fatigue, and tolerate demanding training blocks. However, football is not a simple endurance sport. The decisive moments frequently arise from brief accelerations, rapid changes of direction, high-speed runs into space, and repeated pressing actions.

Consider an academy player who dominates because of early maturation. The scouting report should distinguish current physical advantage from adaptable skill, perception, learning behaviour, and resilience. Long-term development depends on identifying qualities that will still matter when physical differences narrow.

Many football studies are observational. They describe relationships but do not always prove causality. A variable may correlate with success because good teams create the conditions for it, not because the variable alone produces success.

The first touch is a critical bridge between receiving and acting. A good first touch can eliminate pressure, open the next pass, prepare a shot, or change the rhythm of an attack. A poor touch can compress time, invite pressing, and reduce the tactical options available to the team.

In practical terms, running mechanics in football contexts should be understood as part of the larger performance ecology of football. The scientific task is to describe the relevant mechanisms, while the coaching task is to transform that knowledge into training, selection, communication, and

match decisions.

3.2 Acceleration, Braking, and Change of Direction

Football performance depends on repeated high-intensity actions embedded within long periods of lower-intensity movement. Players walk, jog, accelerate, decelerate, sprint, jump, collide, and recover, often within short time windows. The energy cost of the game is therefore not represented adequately by total distance alone; the pattern and timing of intense actions are equally important.

Aerobic fitness supports football because it helps players recover between explosive efforts, maintain technical quality during fatigue, and tolerate demanding training blocks. However, football is not a simple endurance sport. The decisive moments frequently arise from brief accelerations, rapid changes of direction, high-speed runs into space, and repeated pressing actions.

Physiological preparation must respect positional demands. Full-backs and wide midfielders may require repeated high-speed running along corridors of the pitch, centre-backs may face intense accelerations and aerial duels, and central midfielders often accumulate large volumes of medium-intensity movement while constantly adjusting position.

A practical football environment also requires shared language. Coaches may describe a situation tactically, medical staff may describe it through load or risk, analysts may describe it through spatial data, and players may describe it through feel. The interdisciplinary task is to translate these perspectives into a coherent plan.

The first touch is a critical bridge between receiving and acting. A good first touch can eliminate pressure, open the next pass, prepare a shot, or change the rhythm of an attack. A poor touch can compress time, invite pressing, and reduce the tactical options available to the team.

Consider a wide player who repeatedly makes high-speed runs behind the defensive line. A physical report may describe sprint distance, but tactical analysis must ask whether the runs were timed well, whether teammates had passing access, whether the runs stretched the opponent, and whether the player could repeat them after fatigue. The scientific value appears when physical output is connected to tactical effect.

The best evidence-informed practice combines published research, local data, practitioner expertise, and player feedback. None of these sources is sufficient on its own.

Acceleration requires forward lean, effective ground contact, and rapid force production. Deceleration requires the opposite skill: the capacity to brake safely and reorient the body. In football, braking may be as important as sprinting because tactical success often depends on stopping, turning, and acting again before an opponent can respond.

In practical terms, acceleration, braking, and change of direction should be understood as part of the larger performance ecology of football. The scientific task is to describe the relevant mechanisms, while the coaching task is to transform that knowledge into training, selection, communication, and match decisions.

3.3 The Biomechanics of Kicking

Technical skill in football is perception-action skill. A pass is not only a motor pattern but a decision made in relation to pressure, timing, space, and teammate movement. The best players do not merely execute techniques well; they select the appropriate technique earlier and under more difficult conditions.

Motor learning research supports practice environments that include variability, representative pressure, and meaningful information. Repetition remains important, but repetition without context may produce skill that appears clean in drills and fragile in matches. Practice should help players recognize cues, adapt body shape, and execute under realistic constraints.

The first touch is a critical bridge between receiving and acting. A good first touch can eliminate pressure, open the next pass, prepare a shot, or change the rhythm of an attack. A poor touch can compress time, invite pressing, and reduce the tactical options available to the team.

The strongest programmes are iterative. They observe performance, identify problems, design interventions, test them in training, and adjust them according to evidence and player response. This cycle resembles scientific method, but it must operate at the speed of a competitive season.

Biomechanics explains how forces are produced, transferred, absorbed, and redirected during football actions. Running, kicking, heading, landing, and turning are not merely matters of effort; they are coordinated movement solutions. Efficient technique allows the player to produce speed or power while limiting unnecessary stress on joints and soft tissue.

Consider a central midfielder who completes many short passes. A raw passing percentage may look excellent, but the analyst must examine pressure, pass direction, receiving body shape, and whether the passes progressed play. Football data becomes meaningful when it differentiates safe circulation from actions that alter the opponent's defensive structure.

Many football studies are observational. They describe relationships but do not always prove causality. A variable may correlate with success because good teams create the conditions for it, not because the variable alone produces success.

Football injuries emerge from the interaction between tissue capacity, movement demands, fatigue, contact, history, and training load. No single variable explains risk. Prevention therefore requires a system that combines strength, neuromuscular control, appropriate workload progression, recovery, and clear communication between coaches and medical staff.

In practical terms, the biomechanics of kicking should be understood as part of the larger performance ecology of football. The scientific task is to describe the relevant mechanisms, while the coaching task is to transform that knowledge into training, selection, communication, and match decisions.

3.4 Jumping, Heading, and Aerial Duels

Biomechanics explains how forces are produced, transferred, absorbed, and redirected during football actions. Running, kicking, heading, landing, and turning are not merely matters of effort; they are coordinated movement solutions. Efficient technique allows the player to produce speed or power while limiting unnecessary stress on joints and soft tissue.

Acceleration requires forward lean, effective ground contact, and rapid force production. Deceleration requires the opposite skill: the capacity to brake safely and reorient the body. In football, braking may be as important as sprinting because tactical success often depends on stopping, turning, and acting again before an opponent can respond.

Kicking is a whole-body action. The support foot, pelvis, trunk, hip, knee, and ankle contribute to ball speed, accuracy, and trajectory. Technical coaching that focuses only on the striking leg misses the chain of coordination that produces a powerful and controlled strike.

Context should protect practitioners from false certainty. A metric that is useful for one team model may be misleading for another. A player who appears inefficient in one tactical structure may become highly valuable in a different role. Scientific interpretation must therefore remain conditional and specific.

Hamstring injuries are common because football repeatedly exposes players to sprinting, high-speed running, and rapid changes in muscle length. Prevention strategies usually combine sprint exposure, eccentric strength, progressive loading, and careful monitoring after congested match periods.

Consider a defender returning from a hamstring injury. Laboratory strength tests are useful, but they do not fully represent the demands of chasing an attacker, braking sharply, turning under pressure, and clearing the ball while fatigued. Return-to-play should therefore progress from controlled tests to football-specific exposure.

The best evidence-informed practice combines published research, local data, practitioner expertise, and player feedback. None of these sources is sufficient on its own.

Physiological preparation must respect positional demands. Full-backs and wide midfielders may require repeated high-speed running along corridors of the pitch, centre-backs may face intense accelerations and aerial duels, and central midfielders often accumulate large volumes of medium-intensity movement while constantly adjusting position.

In practical terms, jumping, heading, and aerial duels should be understood as part of the larger performance ecology of football. The scientific task is to describe the relevant mechanisms, while the coaching task is to transform that knowledge into training, selection, communication, and match decisions.

3.5 Balance, Stability, and Body Orientation

Football injuries emerge from the interaction between tissue capacity, movement demands, fatigue, contact, history, and training load. No single variable explains risk. Prevention therefore requires a system that combines strength, neuromuscular control, appropriate workload progression, recovery, and clear communication between coaches and medical staff.

Hamstring injuries are common because football repeatedly exposes players to sprinting, high-speed running, and rapid changes in muscle length. Prevention strategies usually combine sprint exposure, eccentric strength, progressive loading, and careful monitoring after congested match periods.

Return-to-play decisions should not be based only on the absence of pain. A player may be pain-free yet not ready for the speed, contact, fatigue, and decision-making demands of competition. The final stage of rehabilitation must recreate football-specific stress in a controlled progression.

For applied work, the central question is not whether the variable can be measured, but whether it changes decisions. A coach, analyst, or sports scientist should ask how the information affects training design, match preparation, player development, or medical risk management. Measurement without decision relevance adds administrative burden rather than performance value.

Physiological preparation must respect positional demands. Full-backs and wide midfielders may require repeated high-speed running along corridors of the pitch, centre-backs may face intense accelerations and aerial duels, and central midfielders often accumulate large volumes of medium-intensity movement while constantly adjusting position.

Consider an academy player who dominates because of early maturation. The scouting report should distinguish current physical advantage from adaptable skill, perception, learning behaviour, and resilience. Long-term development depends on identifying qualities that will still matter when physical differences narrow.

Many football studies are observational. They describe relationships but do not always prove causality. A variable may correlate with success because good teams create the conditions for it, not because the variable alone produces success.

Motor learning research supports practice environments that include variability, representative pressure, and meaningful information. Repetition remains important, but repetition without context may produce skill that appears clean in drills and fragile in matches. Practice should help players recognize cues, adapt body shape, and execute under realistic constraints.

In practical terms, balance, stability, and body orientation should be understood as part of the larger performance ecology of football. The scientific task is to describe the relevant mechanisms, while the coaching task is to transform that knowledge into training, selection, communication, and match decisions.

3.6 Movement Quality and Injury Risk

Football performance depends on repeated high-intensity actions embedded within long periods of lower-intensity movement. Players walk, jog, accelerate, decelerate, sprint, jump, collide, and recover, often within short time windows. The energy cost of the game is therefore not represented adequately by total distance alone; the pattern and timing of intense actions are equally important.

Aerobic fitness supports football because it helps players recover between explosive efforts, maintain technical quality during fatigue, and tolerate demanding training blocks. However, football is not a simple endurance sport. The decisive moments frequently arise from brief accelerations, rapid changes of direction, high-speed runs into space, and repeated pressing actions.

Physiological preparation must respect positional demands. Full-backs and wide midfielders may require repeated high-speed running along corridors of the pitch, centre-backs may face intense accelerations and aerial duels, and central midfielders often accumulate large volumes of medium-intensity movement while constantly adjusting position.

A practical football environment also requires shared language. Coaches may describe a situation tactically, medical staff may describe it through load or risk, analysts may describe it through spatial data, and players may describe it through feel. The interdisciplinary task is to translate these perspectives into a coherent plan.

Technical skill in football is perception-action skill. A pass is not only a motor pattern but a decision made in relation to pressure, timing, space, and teammate movement. The best players do not merely execute techniques well; they select the appropriate technique earlier and under more difficult conditions.

Consider a wide player who repeatedly makes high-speed runs behind the defensive line. A physical report may describe sprint distance, but tactical analysis must ask whether the runs were timed well, whether teammates had passing access, whether the runs stretched the opponent, and whether the player could repeat them after fatigue. The scientific value appears when physical output is connected to tactical effect.

The best evidence-informed practice combines published research, local data, practitioner expertise, and player feedback. None of these sources is sufficient on its own.

Biomechanics explains how forces are produced, transferred, absorbed, and redirected during football actions. Running, kicking, heading, landing, and turning are not merely matters of effort; they are coordinated movement solutions. Efficient technique allows the player to produce speed or power while limiting unnecessary stress on joints and soft tissue.

In practical terms, movement quality and injury risk should be understood as part of the larger performance ecology of football. The scientific task is to describe the relevant mechanisms, while the coaching task is to transform that knowledge into training, selection, communication, and match decisions.

Chapter 4

Technical Skill and Motor Learning

This chapter studies technical skill as an adaptive behaviour. The concern is not only how a movement looks in isolation, but how players perceive information, select actions, and execute technique under pressure.

4.1 First Touch and Perceptual Coupling

Biomechanics explains how forces are produced, transferred, absorbed, and redirected during football actions. Running, kicking, heading, landing, and turning are not merely matters of effort; they are coordinated movement solutions. Efficient technique allows the player to produce speed or power while limiting unnecessary stress on joints and soft tissue.

Acceleration requires forward lean, effective ground contact, and rapid force production. Deceleration requires the opposite skill: the capacity to brake safely and reorient the body. In football, braking may be as important as sprinting because tactical success often depends on stopping, turning, and acting again before an opponent can respond.

Kicking is a whole-body action. The support foot, pelvis, trunk, hip, knee, and ankle contribute to ball speed, accuracy, and trajectory. Technical coaching that focuses only on the striking leg misses the chain of coordination that produces a powerful and controlled strike.

A practical football environment also requires shared language. Coaches may describe a situation tactically, medical staff may describe it through load or risk, analysts may describe it through spatial data, and players may describe it through feel. The interdisciplinary task is to translate these perspectives into a coherent plan.

Team cohesion is not a vague emotional quality. It appears in communication, pressing commitment, recovery runs, role acceptance, and the willingness to act for collective benefit. A team with strong cohesion can execute demanding tactical plans more consistently.

Consider a central midfielder who completes many short passes. A raw passing percentage may look excellent, but the analyst must examine pressure, pass direction, receiving body shape, and whether the passes progressed play. Football data becomes meaningful when it differentiates safe circulation from actions that alter the opponent's defensive structure.

Thresholds for speed zones, workload alerts, and fatigue markers vary across studies and commercial systems. Practitioners should understand how a threshold is defined before comparing numbers across teams, leagues, or seasons.

Tactics organize individual actions into collective behaviour. Formations provide reference points, but they do not describe the full reality of play. A team may defend in one structure, build attacks in another, press in a third, and alter spacing according to the opponent's behaviour.

In practical terms, first touch and perceptual coupling should be understood as part of the larger performance ecology of football. The scientific task is to describe the relevant mechanisms, while the coaching task is to transform that knowledge into training, selection, communication, and match decisions.

4.2 Passing as a Technical and Tactical Action

Decision-making in football is time pressured and information rich. Players must scan before receiving, anticipate movement, recognize patterns, and decide while opponents attempt to hide cues or create deception. Cognitive speed is therefore inseparable from technical speed.

Confidence affects performance because it influences attention, risk tolerance, and emotional control. Under pressure, players may narrow their focus, rush decisions, or avoid demanding actions. Psychological preparation helps players maintain useful attention and commit to appropriate decisions in unstable moments.

Team cohesion is not a vague emotional quality. It appears in communication, pressing commitment, recovery runs, role acceptance, and the willingness to act for collective benefit. A team with strong cohesion can execute demanding tactical plans more consistently.

The strongest programmes are iterative. They observe performance, identify problems, design interventions, test them in training, and adjust them according to evidence and player response. This cycle resembles scientific method, but it must operate at the speed of a competitive season.

Tactics organize individual actions into collective behaviour. Formations provide reference points, but they do not describe the full reality of play. A team may defend in one structure, build attacks in another, press in a third, and alter spacing according to the opponent's behaviour.

Consider a defender returning from a hamstring injury. Laboratory strength tests are useful, but they do not fully represent the demands of chasing an attacker, braking sharply, turning under pressure, and clearing the ball while fatigued. Return-to-play should therefore progress from controlled tests to football-specific exposure.

Research evidence should be read with attention to sample size, playing level, age group, sex, competitive context, and measurement technology. Findings from elite male professionals do not always transfer directly to youth football, amateur environments, or women's football.

The first touch is a critical bridge between receiving and acting. A good first touch can eliminate pressure, open the next pass, prepare a shot, or change the rhythm of an attack. A poor touch can compress time, invite pressing, and reduce the tactical options available to the team.

In practical terms, passing as a technical and tactical action should be understood as part of the larger performance ecology of football. The scientific task is to describe the relevant mechanisms, while the coaching task is to transform that knowledge into training, selection, communication, and match decisions.

4.3 Dribbling, Feinting, and Ball Manipulation

Tactics organize individual actions into collective behaviour. Formations provide reference points, but they do not describe the full reality of play. A team may defend in one structure, build attacks in another, press in a third, and alter spacing according to the opponent's behaviour.

Pressing is not simply running at the ball. Effective pressing requires triggers, cover shadows, distances between lines, timing from supporting players, and a clear plan for what happens if the press is broken. Poorly organized pressing can create large spaces behind the first line of pressure.

Transitions are decisive because teams are temporarily disorganized when possession changes. The first few seconds after losing or winning the ball can determine whether a team creates a chance, prevents a counterattack, or loses control of the match.

Context should protect practitioners from false certainty. A metric that is useful for one team model may be misleading for another. A player who appears inefficient in one tactical structure may become highly valuable in a different role. Scientific interpretation must therefore remain conditional and specific.

Motor learning research supports practice environments that include variability, representative pressure, and meaningful information. Repetition remains important, but repetition without context may produce skill that appears clean in drills and fragile in matches. Practice should help players recognize cues, adapt body shape, and execute under realistic constraints.

Consider an academy player who dominates because of early maturation. The scouting report should distinguish current physical advantage from adaptable skill, perception, learning behaviour, and resilience. Long-term development depends on identifying qualities that will still matter when physical differences narrow.

Thresholds for speed zones, workload alerts, and fatigue markers vary across studies and commercial systems. Practitioners should understand how a threshold is defined before comparing numbers across teams, leagues, or seasons.

Acceleration requires forward lean, effective ground contact, and rapid force production. Deceleration requires the opposite skill: the capacity to brake safely and reorient the body. In football, braking may be as important as sprinting because tactical success often depends on stopping, turning, and acting again before an opponent can respond.

In practical terms, dribbling, feinting, and ball manipulation should be understood as part of the larger performance ecology of football. The scientific task is to describe the relevant mechanisms, while the coaching task is to transform that knowledge into training, selection, communication, and match decisions.

4.4 Shooting Skill Under Pressure

Technical skill in football is perception-action skill. A pass is not only a motor pattern but a decision made in relation to pressure, timing, space, and teammate movement. The best players do not merely execute techniques well; they select the appropriate technique earlier and under more difficult conditions.

Motor learning research supports practice environments that include variability, representative pressure, and meaningful information. Repetition remains important, but repetition without context may produce skill that appears clean in drills and fragile in matches. Practice should help players recognize cues, adapt body shape, and execute under realistic constraints.

The first touch is a critical bridge between receiving and acting. A good first touch can eliminate pressure, open the next pass, prepare a shot, or change the rhythm of an attack. A poor touch can compress time, invite pressing, and reduce the tactical options available to the team.

For applied work, the central question is not whether the variable can be measured, but whether it changes decisions. A coach, analyst, or sports scientist should ask how the information affects training design, match preparation, player development, or medical risk management. Measurement without decision relevance adds administrative burden rather than performance value.

Kicking is a whole-body action. The support foot, pelvis, trunk, hip, knee, and ankle contribute to ball speed, accuracy, and trajectory. Technical coaching that focuses only on the striking leg misses the chain of coordination that produces a powerful and controlled strike.

Consider a wide player who repeatedly makes high-speed runs behind the defensive line. A physical report may describe sprint distance, but tactical analysis must ask whether the runs were timed well, whether teammates had passing access, whether the runs stretched the opponent, and whether the player could repeat them after fatigue. The scientific value appears when physical output is connected to tactical effect.

Research evidence should be read with attention to sample size, playing level, age group, sex, competitive context, and measurement technology. Findings from elite male professionals do not always transfer directly to youth football, amateur environments, or women's football.

Decision-making in football is time pressured and information rich. Players must scan before receiving, anticipate movement, recognize patterns, and decide while opponents attempt to hide cues or create deception. Cognitive speed is therefore inseparable from technical speed.

In practical terms, shooting skill under pressure should be understood as part of the larger performance ecology of football. The scientific task is to describe the relevant mechanisms, while the coaching task is to transform that knowledge into training, selection, communication, and match decisions.

4.5 Practice Design and Skill Acquisition

Biomechanics explains how forces are produced, transferred, absorbed, and redirected during football actions. Running, kicking, heading, landing, and turning are not merely matters of effort; they are coordinated movement solutions. Efficient technique allows the player to produce speed or power while limiting unnecessary stress on joints and soft tissue.

Acceleration requires forward lean, effective ground contact, and rapid force production. Deceleration requires the opposite skill: the capacity to brake safely and reorient the body. In football, braking may be as important as sprinting because tactical success often depends on stopping, turning, and acting again before an opponent can respond.

Kicking is a whole-body action. The support foot, pelvis, trunk, hip, knee, and ankle contribute to ball speed, accuracy, and trajectory. Technical coaching that focuses only on the striking leg misses the chain of coordination that produces a powerful and controlled strike.

A practical football environment also requires shared language. Coaches may describe a situation tactically, medical staff may describe it through load or risk, analysts may describe it through spatial data, and players may describe it through feel. The interdisciplinary task is to translate these perspectives into a coherent plan.

Decision-making in football is time pressured and information rich. Players must scan before receiving, anticipate movement, recognize patterns, and decide while opponents attempt to hide cues or create deception. Cognitive speed is therefore inseparable from technical speed.

Consider a central midfielder who completes many short passes. A raw passing percentage may look excellent, but the analyst must examine pressure, pass direction, receiving body shape, and whether the passes progressed play. Football data becomes meaningful when it differentiates safe circulation from actions that alter the opponent's defensive structure.

Thresholds for speed zones, workload alerts, and fatigue markers vary across studies and commercial systems. Practitioners should understand how a threshold is defined before comparing numbers across teams, leagues, or seasons.

Transitions are decisive because teams are temporarily disorganized when possession changes. The first few seconds after losing or winning the ball can determine whether a team creates a chance, prevents a counterattack, or loses control of the match.

In practical terms, practice design and skill acquisition should be understood as part of the larger performance ecology of football. The scientific task is to describe the relevant mechanisms, while the coaching task is to transform that knowledge into training, selection, communication, and match decisions.

4.6 Feedback, Variability, and Transfer to Match Play

Decision-making in football is time pressured and information rich. Players must scan before receiving, anticipate movement, recognize patterns, and decide while opponents attempt to hide cues or create deception. Cognitive speed is therefore inseparable from technical speed.

Confidence affects performance because it influences attention, risk tolerance, and emotional control. Under pressure, players may narrow their focus, rush decisions, or avoid demanding actions. Psychological preparation helps players maintain useful attention and commit to appropriate decisions in unstable moments.

Team cohesion is not a vague emotional quality. It appears in communication, pressing commitment, recovery runs, role acceptance, and the willingness to act for collective benefit. A team with strong cohesion can execute demanding tactical plans more consistently.

The strongest programmes are iterative. They observe performance, identify problems, design interventions, test them in training, and adjust them according to evidence and player response. This cycle resembles scientific method, but it must operate at the speed of a competitive season.

Pressing is not simply running at the ball. Effective pressing requires triggers, cover shadows, distances between lines, timing from supporting players, and a clear plan for what happens if the press is broken. Poorly organized pressing can create large spaces behind the first line of pressure.

Consider a defender returning from a hamstring injury. Laboratory strength tests are useful, but they do not fully represent the demands of chasing an attacker, braking sharply, turning under pressure, and clearing the ball while fatigued. Return-to-play should therefore progress from controlled tests to football-specific exposure.

Research evidence should be read with attention to sample size, playing level, age group, sex, competitive context, and measurement technology. Findings from elite male professionals do not always transfer directly to youth football, amateur environments, or women's football.

Motor learning research supports practice environments that include variability, representative pressure, and meaningful information. Repetition remains important, but repetition without context may produce skill that appears clean in drills and fragile in matches. Practice should help players recognize cues, adapt body shape, and execute under realistic constraints.

In practical terms, feedback, variability, and transfer to match play should be understood as part of the larger performance ecology of football. The scientific task is to describe the relevant mechanisms, while the coaching task is to transform that knowledge into training, selection, communication, and match decisions.

Chapter 5

Tactical Structures and Collective Behaviour

This chapter examines tactical organization and collective behaviour. Football tactics are dynamic rather than fixed; they describe how teams coordinate space, timing, pressure, and risk across different phases of play.

5.1 Formations as Flexible Reference Systems

Performance analysis converts match events into interpretable information. Event data records actions such as passes, shots, tackles, and carries, while tracking data captures player and ball positions over time. Together they allow analysts to study not only what happened, but also where, when, and under which spatial conditions it happened.

Expected goals models estimate the probability that a shot becomes a goal based on historical patterns and shot characteristics. The metric is useful because it separates chance quality from finishing outcome, but it must be interpreted carefully. Models differ in their inputs, and a single match can be strongly shaped by variance.

Data becomes valuable only when it supports better questions. A dashboard full of numbers can distract a coaching staff if the metrics are disconnected from the team's model of play. Good analysis should clarify decisions, not replace football judgement.

The strongest programmes are iterative. They observe performance, identify problems, design interventions, test them in training, and adjust them according to evidence and player response. This cycle resembles scientific method, but it must operate at the speed of a competitive season.

Football performance depends on repeated high-intensity actions embedded within long periods of lower-intensity movement. Players walk, jog, accelerate, decelerate, sprint, jump, collide, and recover, often within short time windows. The energy cost of the game is therefore not represented adequately by total distance alone; the pattern and timing of intense actions are equally important.

Consider an academy player who dominates because of early maturation. The scouting report should distinguish current physical advantage from adaptable skill, perception, learning behaviour, and resilience. Long-term development depends on identifying qualities that will still matter when physical differences narrow.

The best evidence-informed practice combines published research, local data, practitioner expertise, and player feedback. None of these sources is sufficient on its own.

Confidence affects performance because it influences attention, risk tolerance, and emotional control. Under pressure, players may narrow their focus, rush decisions, or avoid demanding actions. Psychological preparation helps players maintain useful attention and commit to appropriate decisions in unstable moments.

In practical terms, formations as flexible reference systems should be understood as part of the larger performance ecology of football. The scientific task is to describe the relevant mechanisms, while the coaching task is to transform that knowledge into training, selection, communication, and match decisions.

5.2 Attacking Principles and Space Creation

Football performance depends on repeated high-intensity actions embedded within long periods of lower-intensity movement. Players walk, jog, accelerate, decelerate, sprint, jump, collide, and recover, often within short time windows. The energy cost of the game is therefore not represented adequately by total distance alone; the pattern and timing of intense actions are equally important.

Aerobic fitness supports football because it helps players recover between explosive efforts, maintain technical quality during fatigue, and tolerate demanding training blocks. However, football is not a simple endurance sport. The decisive moments frequently arise from brief accelerations, rapid changes of direction, high-speed runs into space, and repeated pressing actions.

Physiological preparation must respect positional demands. Full-backs and wide midfielders may require repeated high-speed running along corridors of the pitch, centre-backs may face intense accelerations and aerial duels, and central midfielders often accumulate large volumes of medium-intensity movement while constantly adjusting position.

Context should protect practitioners from false certainty. A metric that is useful for one team model may be misleading for another. A player who appears inefficient in one tactical structure may become highly valuable in a different role. Scientific interpretation must therefore remain conditional and specific.

Confidence affects performance because it influences attention, risk tolerance, and emotional control. Under pressure, players may narrow their focus, rush decisions, or avoid demanding actions. Psychological preparation helps players maintain useful attention and commit to appropriate decisions in unstable moments.

Consider a wide player who repeatedly makes high-speed runs behind the defensive line. A physical report may describe sprint distance, but tactical analysis must ask whether the runs were timed well, whether teammates had passing access, whether the runs stretched the opponent, and whether the player could repeat them after fatigue. The scientific value appears when physical output is connected to tactical effect.

Many football studies are observational. They describe relationships but do not always prove causality. A variable may correlate with success because good teams create the conditions for it, not because the variable alone produces success.

Tactics organize individual actions into collective behaviour. Formations provide reference points, but they do not describe the full reality of play. A team may defend in one structure, build attacks in another, press in a third, and alter spacing according to the opponent's behaviour.

In practical terms, attacking principles and space creation should be understood as part of the larger performance ecology of football. The scientific task is to describe the relevant mechanisms, while the coaching task is to transform that knowledge into training, selection, communication, and match decisions.

5.3 Defensive Compactness and Pressing

Decision-making in football is time pressured and information rich. Players must scan before receiving, anticipate movement, recognize patterns, and decide while opponents attempt to hide cues or create deception. Cognitive speed is therefore inseparable from technical speed.

Confidence affects performance because it influences attention, risk tolerance, and emotional control. Under pressure, players may narrow their focus, rush decisions, or avoid demanding actions. Psychological preparation helps players maintain useful attention and commit to appropriate decisions in unstable moments.

Team cohesion is not a vague emotional quality. It appears in communication, pressing commitment, recovery runs, role acceptance, and the willingness to act for collective benefit. A team with strong cohesion can execute demanding tactical plans more consistently.

For applied work, the central question is not whether the variable can be measured, but whether it changes decisions. A coach, analyst, or sports scientist should ask how the information affects training design, match preparation, player development, or medical risk management. Measurement without decision relevance adds administrative burden rather than performance value.

Transitions are decisive because teams are temporarily disorganized when possession changes. The first few seconds after losing or winning the ball can determine whether a team creates a chance, prevents a counterattack, or loses control of the match.

Consider a central midfielder who completes many short passes. A raw passing percentage may look excellent, but the analyst must examine pressure, pass direction, receiving body shape, and whether the passes progressed play. Football data becomes meaningful when it differentiates safe circulation from actions that alter the opponent's defensive structure.

The best evidence-informed practice combines published research, local data, practitioner expertise, and player feedback. None of these sources is sufficient on its own.

Data becomes valuable only when it supports better questions. A dashboard full of numbers can distract a coaching staff if the metrics are disconnected from the team's model of play. Good analysis should clarify decisions, not replace football judgement.

In practical terms, defensive compactness and pressing should be understood as part of the larger performance ecology of football. The scientific task is to describe the relevant mechanisms, while the coaching task is to transform that knowledge into training, selection, communication, and match decisions.

5.4 Transitions and Counter-Pressing

Tactics organize individual actions into collective behaviour. Formations provide reference points, but they do not describe the full reality of play. A team may defend in one structure, build attacks in another, press in a third, and alter spacing according to the opponent's behaviour.

Pressing is not simply running at the ball. Effective pressing requires triggers, cover shadows, distances between lines, timing from supporting players, and a clear plan for what happens if the press is broken. Poorly organized pressing can create large spaces behind the first line of pressure.

Transitions are decisive because teams are temporarily disorganized when possession changes. The first few seconds after losing or winning the ball can determine whether a team creates a chance, prevents a counterattack, or loses control of the match.

A practical football environment also requires shared language. Coaches may describe a situation tactically, medical staff may describe it through load or risk, analysts may describe it through spatial data, and players may describe it through feel. The interdisciplinary task is to translate these perspectives into a coherent plan.

Performance analysis converts match events into interpretable information. Event data records actions such as passes, shots, tackles, and carries, while tracking data captures player and ball positions over time. Together they allow analysts to study not only what happened, but also where, when, and under which spatial conditions it happened.

Consider a defender returning from a hamstring injury. Laboratory strength tests are useful, but they do not fully represent the demands of chasing an attacker, braking sharply, turning under pressure, and clearing the ball while fatigued. Return-to-play should therefore progress from controlled tests to football-specific exposure.

Many football studies are observational. They describe relationships but do not always prove causality. A variable may correlate with success because good teams create the conditions for it, not because the variable alone produces success.

Aerobic fitness supports football because it helps players recover between explosive efforts, maintain technical quality during fatigue, and tolerate demanding training blocks. However, football is not a simple endurance sport. The decisive moments frequently arise from brief accelerations, rapid changes of direction, high-speed runs into space, and repeated pressing actions.

In practical terms, transitions and counter-pressing should be understood as part of the larger performance ecology of football. The scientific task is to describe the relevant mechanisms, while the coaching task is to transform that knowledge into training, selection, communication, and match decisions.

5.5 Set Pieces as Structured Tactical Events

Performance analysis converts match events into interpretable information. Event data records actions such as passes, shots, tackles, and carries, while tracking data captures player and ball positions over time. Together they allow analysts to study not only what happened, but also where, when, and under which spatial conditions it happened.

Expected goals models estimate the probability that a shot becomes a goal based on historical patterns and shot characteristics. The metric is useful because it separates chance quality from finishing outcome, but it must be interpreted carefully. Models differ in their inputs, and a single match can be strongly shaped by variance.

Data becomes valuable only when it supports better questions. A dashboard full of numbers can distract a coaching staff if the metrics are disconnected from the team's model of play. Good analysis should clarify decisions, not replace football judgement.

The strongest programmes are iterative. They observe performance, identify problems, design interventions, test them in training, and adjust them according to evidence and player response. This cycle resembles scientific method, but it must operate at the speed of a competitive season.

Aerobic fitness supports football because it helps players recover between explosive efforts, maintain technical quality during fatigue, and tolerate demanding training blocks. However, football is not a simple endurance sport. The decisive moments frequently arise from brief accelerations, rapid changes of direction, high-speed runs into space, and repeated pressing actions.

Consider an academy player who dominates because of early maturation. The scouting report should distinguish current physical advantage from adaptable skill, perception, learning behaviour, and resilience. Long-term development depends on identifying qualities that will still matter when physical differences narrow.

The best evidence-informed practice combines published research, local data, practitioner expertise, and player feedback. None of these sources is sufficient on its own.

Decision-making in football is time pressured and information rich. Players must scan before receiving, anticipate movement, recognize patterns, and decide while opponents attempt to hide cues or create deception. Cognitive speed is therefore inseparable from technical speed.

In practical terms, set pieces as structured tactical events should be understood as part of the larger performance ecology of football. The scientific task is to describe the relevant mechanisms, while the coaching task is to transform that knowledge into training, selection, communication, and match decisions.

5.6 Tactical Periodization and Team Identity

Football performance depends on repeated high-intensity actions embedded within long periods of lower-intensity movement. Players walk, jog, accelerate, decelerate, sprint, jump, collide, and recover, often within short time windows. The energy cost of the game is therefore not represented adequately by total distance alone; the pattern and timing of intense actions are equally important.

Aerobic fitness supports football because it helps players recover between explosive efforts, maintain technical quality during fatigue, and tolerate demanding training blocks. However, football is not a simple endurance sport. The decisive moments frequently arise from brief accelerations, rapid changes of direction, high-speed runs into space, and repeated pressing actions.

Physiological preparation must respect positional demands. Full-backs and wide midfielders may require repeated high-speed running along corridors of the pitch, centre-backs may face intense accelerations and aerial duels, and central midfielders often accumulate large volumes of medium-intensity movement while constantly adjusting position.

Context should protect practitioners from false certainty. A metric that is useful for one team model may be misleading for another. A player who appears inefficient in one tactical structure may become highly valuable in a different role. Scientific interpretation must therefore remain conditional and specific.

Team cohesion is not a vague emotional quality. It appears in communication, pressing commitment, recovery runs, role acceptance, and the willingness to act for collective benefit. A team with strong cohesion can execute demanding tactical plans more consistently.

Consider a wide player who repeatedly makes high-speed runs behind the defensive line. A physical report may describe sprint distance, but tactical analysis must ask whether the runs were timed well, whether teammates had passing access, whether the runs stretched the opponent, and whether the player could repeat them after fatigue. The scientific value appears when physical output is connected to tactical effect.

Many football studies are observational. They describe relationships but do not always prove causality. A variable may correlate with success because good teams create the conditions for it, not because the variable alone produces success.

Transitions are decisive because teams are temporarily disorganized when possession changes. The first few seconds after losing or winning the ball can determine whether a team creates a chance, prevents a counterattack, or loses control of the match.

In practical terms, tactical periodization and team identity should be understood as part of the larger performance ecology of football. The scientific task is to describe the relevant mechanisms, while the coaching task is to transform that knowledge into training, selection, communication, and

match decisions.

Chapter 6

Performance Analysis and Football Data

This chapter explains how data and analysis support modern football. It covers the logic of event data, tracking data, expected goals, network analysis, and the interpretation problems that arise when complex play is reduced to numbers.

6.1 Event Data, Tracking Data, and Video Analysis

Tactics organize individual actions into collective behaviour. Formations provide reference points, but they do not describe the full reality of play. A team may defend in one structure, build attacks in another, press in a third, and alter spacing according to the opponent's behaviour.

Pressing is not simply running at the ball. Effective pressing requires triggers, cover shadows, distances between lines, timing from supporting players, and a clear plan for what happens if the press is broken. Poorly organized pressing can create large spaces behind the first line of pressure.

Transitions are decisive because teams are temporarily disorganized when possession changes. The first few seconds after losing or winning the ball can determine whether a team creates a chance, prevents a counterattack, or loses control of the match.

Context should protect practitioners from false certainty. A metric that is useful for one team model may be misleading for another. A player who appears inefficient in one tactical structure may become highly valuable in a different role. Scientific interpretation must therefore remain conditional and specific.

Artificial intelligence can support scouting, opposition analysis, injury-risk modelling, and individualized training design. However, algorithmic outputs are not neutral. They depend on data quality, modelling assumptions, and organizational incentives. Clubs must understand both the strengths and limitations of the systems they use.

Consider a central midfielder who completes many short passes. A raw passing percentage may look excellent, but the analyst must examine pressure, pass direction, receiving body shape, and whether the passes progressed play. Football data becomes meaningful when it differentiates safe circulation from actions that alter the opponent's defensive structure.

Research evidence should be read with attention to sample size, playing level, age group, sex, competitive context, and measurement technology. Findings from elite male professionals do not always transfer directly to youth football, amateur environments, or women's football.

The scientific study of football must balance reduction and integration. Reduction allows the researcher to isolate sprint speed, joint angle, oxygen uptake, or passing frequency. Integration asks how these variables function inside a match. Both approaches are necessary; the danger lies in treating an isolated variable as if it explains performance by itself.

In practical terms, event data, tracking data, and video analysis should be understood as part of the larger performance ecology of football. The scientific task is to describe the relevant mechanisms, while the coaching task is to transform that knowledge into training, selection, communication, and match decisions.

6.2 Expected Goals and Chance Quality

Technology will continue to change how football is observed, trained, and governed. Wearables, optical tracking, automated tagging, and artificial intelligence can reveal patterns that are difficult to see manually. Yet technology must remain subordinate to ethical practice, player welfare, and football intelligence.

Artificial intelligence can support scouting, opposition analysis, injury-risk modelling, and individualized training design. However, algorithmic outputs are not neutral. They depend on data quality, modelling assumptions, and organizational incentives. Clubs must understand both the strengths and limitations of the systems they use.

The future coach will increasingly operate as a scientist-practitioner: someone able to interpret evidence, test ideas, collaborate with specialists, and translate complex information into clear training tasks. The best environments will combine rigorous analysis with human understanding.

For applied work, the central question is not whether the variable can be measured, but whether it changes decisions. A coach, analyst, or sports scientist should ask how the information affects training design, match preparation, player development, or medical risk management. Measurement without decision relevance adds administrative burden rather than performance value.

The scientific study of football must balance reduction and integration. Reduction allows the researcher to isolate sprint speed, joint angle, oxygen uptake, or passing frequency. Integration asks how these variables function inside a match. Both approaches are necessary; the danger lies in treating an isolated variable as if it explains performance by itself.

Consider a defender returning from a hamstring injury. Laboratory strength tests are useful, but they do not fully represent the demands of chasing an attacker, braking sharply, turning under pressure, and clearing the ball while fatigued. Return-to-play should therefore progress from controlled tests to football-specific exposure.

Thresholds for speed zones, workload alerts, and fatigue markers vary across studies and commercial systems. Practitioners should understand how a threshold is defined before comparing numbers across teams, leagues, or seasons.

Expected goals models estimate the probability that a shot becomes a goal based on historical patterns and shot characteristics. The metric is useful because it separates chance quality from finishing outcome, but it must be interpreted carefully. Models differ in their inputs, and a single match can be strongly shaped by variance.

In practical terms, expected goals and chance quality should be understood as part of the larger performance ecology of football. The scientific task is to describe the relevant mechanisms, while the coaching task is to transform that knowledge into training, selection, communication, and match decisions.

6.3 Passing Networks and Positional Relationships

Football is a complex adaptive sport rather than a closed laboratory task. The same pass, sprint, tackle, or shot can have different value depending on the score, the opponent's shape, the phase of play, the time in the match, and the individual qualities of the players involved. Scientific analysis is therefore most useful when it respects the ecology of the game and connects measurement to tactical meaning.

A match is produced by interacting constraints: rules, pitch dimensions, weather, fatigue, the referee's interpretation, coaching strategy, crowd pressure, and the changing distribution of space. These constraints do not operate separately. They continuously reshape the options available to each player and determine whether a technical action is easy, risky, delayed, or impossible.

The scientific study of football must balance reduction and integration. Reduction allows the researcher to isolate sprint speed, joint angle, oxygen uptake, or passing frequency. Integration asks how these variables function inside a match. Both approaches are necessary; the danger lies in treating an isolated variable as if it explains performance by itself.

A practical football environment also requires shared language. Coaches may describe a situation tactically, medical staff may describe it through load or risk, analysts may describe it through spatial data, and players may describe it through feel. The interdisciplinary task is to translate these perspectives into a coherent plan.

Performance analysis converts match events into interpretable information. Event data records actions such as passes, shots, tackles, and carries, while tracking data captures player and ball positions over time. Together they allow analysts to study not only what happened, but also where, when, and under which spatial conditions it happened.

Consider an academy player who dominates because of early maturation. The scouting report should distinguish current physical advantage from adaptable skill, perception, learning behaviour, and resilience. Long-term development depends on identifying qualities that will still matter when physical differences narrow.

Research evidence should be read with attention to sample size, playing level, age group, sex, competitive context, and measurement technology. Findings from elite male professionals do not always transfer directly to youth football, amateur environments, or women's football.

Tactics organize individual actions into collective behaviour. Formations provide reference points, but they do not describe the full reality of play. A team may defend in one structure, build attacks in another, press in a third, and alter spacing according to the opponent's behaviour.

In practical terms, passing networks and positional relationships should be understood as part of the larger performance ecology of football. The scientific task is to describe the relevant mechanisms, while the coaching task is to transform that knowledge into training, selection, communication, and match decisions.

6.4 Physical Metrics in Tactical Context

Performance analysis converts match events into interpretable information. Event data records actions such as passes, shots, tackles, and carries, while tracking data captures player and ball positions over time. Together they allow analysts to study not only what happened, but also where, when, and under which spatial conditions it happened.

Expected goals models estimate the probability that a shot becomes a goal based on historical patterns and shot characteristics. The metric is useful because it separates chance quality from finishing outcome, but it must be interpreted carefully. Models differ in their inputs, and a single match can be strongly shaped by variance.

Data becomes valuable only when it supports better questions. A dashboard full of numbers can distract a coaching staff if the metrics are disconnected from the team's model of play. Good analysis should clarify decisions, not replace football judgement.

The strongest programmes are iterative. They observe performance, identify problems, design interventions, test them in training, and adjust them according to evidence and player response. This cycle resembles scientific method, but it must operate at the speed of a competitive season.

Pressing is not simply running at the ball. Effective pressing requires triggers, cover shadows, distances between lines, timing from supporting players, and a clear plan for what happens if the press is broken. Poorly organized pressing can create large spaces behind the first line of pressure.

Consider a wide player who repeatedly makes high-speed runs behind the defensive line. A physical report may describe sprint distance, but tactical analysis must ask whether the runs were timed well, whether teammates had passing access, whether the runs stretched the opponent, and whether the player could repeat them after fatigue. The scientific value appears when physical output is connected to tactical effect.

Thresholds for speed zones, workload alerts, and fatigue markers vary across studies and commercial systems. Practitioners should understand how a threshold is defined before comparing numbers across teams, leagues, or seasons.

The future coach will increasingly operate as a scientist-practitioner: someone able to interpret evidence, test ideas, collaborate with specialists, and translate complex information into clear training tasks. The best environments will combine rigorous analysis with human understanding.

In practical terms, physical metrics in tactical context should be understood as part of the larger performance ecology of football. The scientific task is to describe the relevant mechanisms, while the coaching task is to transform that knowledge into training, selection, communication, and match decisions.

6.5 From Descriptive Statistics to Decision Support

Tactics organize individual actions into collective behaviour. Formations provide reference points, but they do not describe the full reality of play. A team may defend in one structure, build attacks in another, press in a third, and alter spacing according to the opponent's behaviour.

Pressing is not simply running at the ball. Effective pressing requires triggers, cover shadows, distances between lines, timing from supporting players, and a clear plan for what happens if the press is broken. Poorly organized pressing can create large spaces behind the first line of pressure.

Transitions are decisive because teams are temporarily disorganized when possession changes. The first few seconds after losing or winning the ball can determine whether a team creates a chance, prevents a counterattack, or loses control of the match.

Context should protect practitioners from false certainty. A metric that is useful for one team model may be misleading for another. A player who appears inefficient in one tactical structure may become highly valuable in a different role. Scientific interpretation must therefore remain conditional and specific.

The future coach will increasingly operate as a scientist-practitioner: someone able to interpret evidence, test ideas, collaborate with specialists, and translate complex information into clear training tasks. The best environments will combine rigorous analysis with human understanding.

Consider a central midfielder who completes many short passes. A raw passing percentage may look excellent, but the analyst must examine pressure, pass direction, receiving body shape, and whether the passes progressed play. Football data becomes meaningful when it differentiates safe circulation from actions that alter the opponent's defensive structure.

Research evidence should be read with attention to sample size, playing level, age group, sex, competitive context, and measurement technology. Findings from elite male professionals do not always transfer directly to youth football, amateur environments, or women's football.

A match is produced by interacting constraints: rules, pitch dimensions, weather, fatigue, the referee's interpretation, coaching strategy, crowd pressure, and the changing distribution of space. These constraints do not operate separately. They continuously reshape the options available to each player and determine whether a technical action is easy, risky, delayed, or impossible.

In practical terms, from descriptive statistics to decision support should be understood as part of the larger performance ecology of football. The scientific task is to describe the relevant mechanisms, while the coaching task is to transform that knowledge into training, selection, communication, and match decisions.

6.6 Limits, Biases, and Ethical Use of Data

Technology will continue to change how football is observed, trained, and governed. Wearables, optical tracking, automated tagging, and artificial intelligence can reveal patterns that are difficult to see manually. Yet technology must remain subordinate to ethical practice, player welfare, and football intelligence.

Artificial intelligence can support scouting, opposition analysis, injury-risk modelling, and individualized training design. However, algorithmic outputs are not neutral. They depend on data quality, modelling assumptions, and organizational incentives. Clubs must understand both the strengths and limitations of the systems they use.

The future coach will increasingly operate as a scientist-practitioner: someone able to interpret evidence, test ideas, collaborate with specialists, and translate complex information into clear training tasks. The best environments will combine rigorous analysis with human understanding.

For applied work, the central question is not whether the variable can be measured, but whether it changes decisions. A coach, analyst, or sports scientist should ask how the information affects training design, match preparation, player development, or medical risk management. Measurement without decision relevance adds administrative burden rather than performance value.

Football is a complex adaptive sport rather than a closed laboratory task. The same pass, sprint, tackle, or shot can have different value depending on the score, the opponent's shape, the phase of play, the time in the match, and the individual qualities of the players involved. Scientific analysis is therefore most useful when it respects the ecology of the game and connects measurement to tactical meaning.

Consider a defender returning from a hamstring injury. Laboratory strength tests are useful, but they do not fully represent the demands of chasing an attacker, braking sharply, turning under pressure, and clearing the ball while fatigued. Return-to-play should therefore progress from controlled tests to football-specific exposure.

Thresholds for speed zones, workload alerts, and fatigue markers vary across studies and commercial systems. Practitioners should understand how a threshold is defined before comparing numbers across teams, leagues, or seasons.

Performance analysis converts match events into interpretable information. Event data records actions such as passes, shots, tackles, and carries, while tracking data captures player and ball positions over time. Together they allow analysts to study not only what happened, but also where, when, and under which spatial conditions it happened.

In practical terms, limits, biases, and ethical use of data should be understood as part of the larger performance ecology of football. The scientific task is to describe the relevant mechanisms, while the coaching task is to transform that knowledge into training, selection, communication, and match decisions.

Chapter 7

Training Methodology and Periodization

This chapter connects training theory to football practice. It considers weekly planning, small-sided games, strength and speed work, recovery, and the challenge of balancing adaptation with freshness during competition.

7.1 The Weekly Microcycle

Technical skill in football is perception-action skill. A pass is not only a motor pattern but a decision made in relation to pressure, timing, space, and teammate movement. The best players do not merely execute techniques well; they select the appropriate technique earlier and under more difficult conditions.

Motor learning research supports practice environments that include variability, representative pressure, and meaningful information. Repetition remains important, but repetition without context may produce skill that appears clean in drills and fragile in matches. Practice should help players recognize cues, adapt body shape, and execute under realistic constraints.

The first touch is a critical bridge between receiving and acting. A good first touch can eliminate pressure, open the next pass, prepare a shot, or change the rhythm of an attack. A poor touch can compress time, invite pressing, and reduce the tactical options available to the team.

For applied work, the central question is not whether the variable can be measured, but whether it changes decisions. A coach, analyst, or sports scientist should ask how the

information affects training design, match preparation, player development, or medical risk management. Measurement without decision relevance adds administrative burden rather than performance value.

Return-to-play decisions should not be based only on the absence of pain. A player may be pain-free yet not ready for the speed, contact, fatigue, and decision-making demands of competition. The final stage of rehabilitation must recreate football-specific stress in a controlled progression.

Consider an academy player who dominates because of early maturation. The scouting report should distinguish current physical advantage from adaptable skill, perception, learning behaviour, and resilience. Long-term development depends on identifying qualities that will still matter when physical differences narrow.

Many football studies are observational. They describe relationships but do not always prove causality. A variable may correlate with success because good teams create the conditions for it, not because the variable alone produces success.

Tactics organize individual actions into collective behaviour. Formations provide reference points, but they do not describe the full reality of play. A team may defend in one structure, build attacks in another, press in a third, and alter spacing according to the opponent's behaviour.

In practical terms, the weekly microcycle should be understood as part of the larger performance ecology of football. The scientific task is to describe the relevant mechanisms, while the coaching task is to transform that knowledge into training, selection, communication, and match decisions.

7.2 Small-Sided Games as Integrated Training

Football injuries emerge from the interaction between tissue capacity, movement demands, fatigue, contact, history, and training load. No single variable explains risk. Prevention therefore requires a system that combines strength, neuromuscular control, appropriate workload progression, recovery, and clear communication between coaches and medical staff.

Hamstring injuries are common because football repeatedly exposes players to sprinting, high-speed running, and rapid changes in muscle length. Prevention strategies usually combine sprint exposure, eccentric strength, progressive loading, and careful monitoring after congested match periods.

Return-to-play decisions should not be based only on the absence of pain. A player may be pain-free yet not ready for the speed, contact, fatigue, and decision-making demands of competition. The final stage of rehabilitation must recreate football-specific stress in a controlled progression.

A practical football environment also requires shared language. Coaches may describe a situation tactically, medical staff may describe it through load or risk, analysts may describe it through spatial data, and players may describe it through feel. The interdisciplinary task is to translate these perspectives into a coherent plan.

Tactics organize individual actions into collective behaviour. Formations provide reference points, but they do not describe the full reality of play. A team may defend in one structure, build attacks in another, press in a third, and alter spacing according to the opponent's behaviour.

Consider a wide player who repeatedly makes high-speed runs behind the defensive line. A physical report may describe sprint distance, but tactical analysis must ask whether the runs were timed well, whether teammates had passing access, whether the runs stretched the opponent, and whether the player could repeat them after fatigue. The scientific value appears when physical output is connected to tactical effect.

The best evidence-informed practice combines published research, local data, practitioner expertise, and player feedback. None of these sources is sufficient on its own.

Physiological preparation must respect positional demands. Full-backs and wide midfielders may require repeated high-speed running along corridors of the pitch, centre-backs may face intense accelerations and aerial duels, and central midfielders often accumulate large volumes of medium-intensity movement while constantly adjusting position.

In practical terms, small-sided games as integrated training should be understood as part of the larger performance ecology of football. The scientific task is to describe the relevant mechanisms, while the coaching task is to transform that knowledge into training, selection, communication, and match decisions.

7.3 Strength and Power Development

Tactics organize individual actions into collective behaviour. Formations provide reference points, but they do not describe the full reality of play. A team may defend in one structure, build attacks in another, press in a third, and alter spacing according to the opponent's behaviour.

Pressing is not simply running at the ball. Effective pressing requires triggers, cover shadows, distances between lines, timing from supporting players, and a clear plan for what happens if the press is broken. Poorly organized pressing can create large spaces behind the first line of pressure.

Transitions are decisive because teams are temporarily disorganized when possession changes. The first few seconds after losing or winning the ball can determine whether a team creates a chance, prevents a counterattack, or loses control of the match.

The strongest programmes are iterative. They observe performance, identify problems, design interventions, test them in training, and adjust them according to evidence and player response. This cycle resembles scientific method, but it must operate at the speed of a competitive season.

Aerobic fitness supports football because it helps players recover between explosive efforts, maintain technical quality during fatigue, and tolerate demanding training blocks. However, football is not a simple endurance sport. The decisive moments frequently arise from brief accelerations, rapid changes of direction, high-speed runs into space, and repeated pressing actions.

Consider a central midfielder who completes many short passes. A raw passing percentage may look excellent, but the analyst must examine pressure, pass direction, receiving body shape, and whether the passes progressed play. Football data becomes meaningful when it differentiates safe circulation from actions that alter the opponent's defensive structure.

Many football studies are observational. They describe relationships but do not always prove causality. A variable may correlate with success because good teams create the conditions for it, not because the variable alone produces success.

Motor learning research supports practice environments that include variability, representative pressure, and meaningful information. Repetition remains important, but repetition without context may produce skill that appears clean in drills and fragile in matches. Practice should help players recognize cues, adapt body shape, and execute under realistic constraints.

In practical terms, strength and power development should be understood as part of the larger performance ecology of football. The scientific task is to describe the relevant mechanisms, while the coaching task is to transform that knowledge into training, selection, communication, and match decisions.

7.4 Speed, Agility, and Repeated Sprint Ability

Football performance depends on repeated high-intensity actions embedded within long periods of lower-intensity movement. Players walk, jog, accelerate, decelerate, sprint, jump, collide, and recover, often within short time windows. The energy cost of the game is therefore not represented adequately by total distance alone; the pattern and timing of intense actions are equally important.

Aerobic fitness supports football because it helps players recover between explosive efforts, maintain technical quality during fatigue, and tolerate demanding training blocks. However, football is not a simple endurance sport. The decisive moments frequently arise from brief accelerations, rapid changes of direction, high-speed runs into space, and repeated pressing actions.

Physiological preparation must respect positional demands. Full-backs and wide midfielders may require repeated high-speed running along corridors of the pitch, centre-backs may face intense accelerations and aerial duels, and central midfielders often accumulate large volumes of medium-intensity movement while constantly adjusting position.

Context should protect practitioners from false certainty. A metric that is useful for one team model may be misleading for another. A player who appears inefficient in one tactical structure may become highly valuable in a different role. Scientific interpretation must therefore remain conditional and specific.

The first touch is a critical bridge between receiving and acting. A good first touch can eliminate pressure, open the next pass, prepare a shot, or change the rhythm of an attack. A poor touch can compress time, invite pressing, and reduce the tactical options available to the team.

Consider a defender returning from a hamstring injury. Laboratory strength tests are useful, but they do not fully represent the demands of chasing an attacker, braking sharply, turning under pressure, and clearing the ball while fatigued. Return-to-play should therefore progress from controlled tests to football-specific exposure.

The best evidence-informed practice combines published research, local data, practitioner expertise, and player feedback. None of these sources is sufficient on its own.

Football injuries emerge from the interaction between tissue capacity, movement demands, fatigue, contact, history, and training load. No single variable explains risk. Prevention therefore requires a system that combines strength, neuromuscular control, appropriate workload progression, recovery, and clear communication between coaches and medical staff.

In practical terms, speed, agility, and repeated sprint ability should be understood as part of the larger performance ecology of football. The scientific task is to describe the relevant mechanisms, while the coaching task is to transform that knowledge into training, selection, communication, and match decisions.

7.5 Recovery, Nutrition, and Adaptation

Technical skill in football is perception-action skill. A pass is not only a motor pattern but a decision made in relation to pressure, timing, space, and teammate movement. The best players do not merely execute techniques well; they select the appropriate technique earlier and under more difficult conditions.

Motor learning research supports practice environments that include variability, representative pressure, and meaningful information. Repetition remains important, but repetition without context may produce skill that appears clean in drills and fragile in matches. Practice should help players recognize cues, adapt body shape, and execute under realistic constraints.

The first touch is a critical bridge between receiving and acting. A good first touch can eliminate pressure, open the next pass, prepare a shot, or change the rhythm of an attack. A poor touch can compress time, invite pressing, and reduce the tactical options available to the team.

For applied work, the central question is not whether the variable can be measured, but whether it changes decisions. A coach, analyst, or sports scientist should ask how the information affects training design, match preparation, player development, or medical risk management. Measurement without decision relevance adds administrative burden rather than performance value.

Football injuries emerge from the interaction between tissue capacity, movement demands, fatigue, contact, history, and training load. No single variable explains risk. Prevention therefore requires a system that combines strength, neuromuscular control, appropriate workload progression, recovery, and clear communication between coaches and medical staff.

Consider an academy player who dominates because of early maturation. The scouting report should distinguish current physical advantage from adaptable skill, perception, learning behaviour, and resilience. Long-term development depends on identifying qualities that will still matter when physical differences narrow.

Many football studies are observational. They describe relationships but do not always prove causality. A variable may correlate with success because good teams create the conditions for it, not because the variable alone produces success.

Transitions are decisive because teams are temporarily disorganized when possession changes. The first few seconds after losing or winning the ball can determine whether a team creates a chance, prevents a counterattack, or loses control of the match.

In practical terms, recovery, nutrition, and adaptation should be understood as part of the larger performance ecology of football. The scientific task is to describe the relevant mechanisms, while the coaching task is to transform that knowledge into training, selection, communication, and match decisions.

7.6 Managing Match Congestion

Football injuries emerge from the interaction between tissue capacity, movement demands, fatigue, contact, history, and training load. No single variable explains risk. Prevention therefore requires a system that combines strength, neuromuscular control, appropriate workload progression, recovery, and clear communication between coaches and medical staff.

Hamstring injuries are common because football repeatedly exposes players to sprinting, high-speed running, and rapid changes in muscle length. Prevention strategies usually combine sprint exposure, eccentric strength, progressive loading, and careful monitoring after congested match periods.

Return-to-play decisions should not be based only on the absence of pain. A player may be pain-free yet not ready for the speed, contact, fatigue, and decision-making demands of competition. The final stage of rehabilitation must recreate football-specific stress in a controlled progression.

A practical football environment also requires shared language. Coaches may describe a situation tactically, medical staff may describe it through load or risk, analysts may describe it through spatial data, and players may describe it through feel. The interdisciplinary task is to translate these perspectives into a coherent plan.

Pressing is not simply running at the ball. Effective pressing requires triggers, cover shadows, distances between lines, timing from supporting players, and a clear plan for what happens if the press is broken. Poorly organized pressing can create large spaces behind the first line of pressure.

Consider a wide player who repeatedly makes high-speed runs behind the defensive line. A physical report may describe sprint distance, but tactical analysis must ask whether the runs were timed well, whether teammates had passing access, whether the runs stretched the opponent, and whether the player could repeat them after fatigue. The scientific value appears when physical output is connected to tactical effect.

The best evidence-informed practice combines published research, local data, practitioner expertise, and player feedback. None of these sources is sufficient on its own.

Aerobic fitness supports football because it helps players recover between explosive efforts, maintain technical quality during fatigue, and tolerate demanding training blocks. However, football is not a simple endurance sport. The decisive moments frequently arise from brief accelerations, rapid changes of direction, high-speed runs into space, and repeated pressing actions.

In practical terms, managing match congestion should be understood as part of the larger performance ecology of football. The scientific task is to describe the relevant mechanisms, while the coaching task is to transform that knowledge into training, selection, communication, and match decisions.

Chapter 8

Injury Prevention and Sports Medicine

This chapter focuses on injury prevention, rehabilitation, and medical decision-making. Football medicine must deal with high-speed running, contact, fatigue, recurrence risk, and the pressure to return players quickly.

8.1 Common Injuries in Football

Biomechanics explains how forces are produced, transferred, absorbed, and redirected during football actions. Running, kicking, heading, landing, and turning are not merely matters of effort; they are coordinated movement solutions. Efficient technique allows the player to produce speed or power while limiting unnecessary stress on joints and soft tissue.

Acceleration requires forward lean, effective ground contact, and rapid force production. Deceleration requires the opposite skill: the capacity to brake safely and reorient the body. In football, braking may be as important as sprinting because tactical success often depends on stopping, turning, and acting again before an opponent can respond.

Kicking is a whole-body action. The support foot, pelvis, trunk, hip, knee, and ankle contribute to ball speed, accuracy, and trajectory. Technical coaching that focuses only on the striking leg misses the chain of coordination that produces a powerful and controlled strike.

A practical football environment also requires shared language. Coaches may describe a situation tactically, medical staff may describe it through load or risk, analysts may describe it through spatial data, and players may describe it through feel. The interdisciplinary task is to translate these perspectives into a coherent plan.

Football performance depends on repeated high-intensity actions embedded within long periods of lower-intensity movement. Players walk, jog, accelerate, decelerate, sprint, jump, collide, and recover, often within short time windows. The energy cost of the game is therefore not represented adequately by total distance alone; the pattern and timing of intense actions are equally important.

Consider a central midfielder who completes many short passes. A raw passing percentage may look excellent, but the analyst must examine pressure, pass direction, receiving body shape, and whether the passes progressed play. Football data becomes meaningful when it differentiates safe circulation from actions that alter the opponent's defensive structure.

Thresholds for speed zones, workload alerts, and fatigue markers vary across studies and commercial systems. Practitioners should understand how a threshold is defined before comparing numbers across teams, leagues, or seasons.

Confidence affects performance because it influences attention, risk tolerance, and emotional control. Under pressure, players may narrow their focus, rush decisions, or avoid demanding actions. Psychological preparation helps players maintain useful attention and commit to appropriate decisions in unstable moments.

In practical terms, common injuries in football should be understood as part of the larger performance ecology of football. The scientific task is to describe the relevant mechanisms, while the coaching task is to transform that knowledge into training, selection, communication, and match decisions.

8.2 Hamstring and Groin Injury Mechanisms

Football performance depends on repeated high-intensity actions embedded within long periods of lower-intensity movement. Players walk, jog, accelerate, decelerate, sprint, jump, collide, and recover, often within short time windows. The energy cost of the game is therefore not represented adequately by total distance alone; the pattern and timing of intense actions are equally important.

Aerobic fitness supports football because it helps players recover between explosive efforts, maintain technical quality during fatigue, and tolerate demanding training blocks. However, football is not a simple endurance sport. The decisive moments frequently arise from brief accelerations, rapid changes of direction, high-speed runs into space, and repeated pressing actions.

Physiological preparation must respect positional demands. Full-backs and wide midfielders may require repeated high-speed running along corridors of the pitch, centre-backs may face intense accelerations and aerial duels, and central midfielders often accumulate large volumes of medium-intensity movement while constantly adjusting position.

The strongest programmes are iterative. They observe performance, identify problems, design interventions, test them in training, and adjust them according to evidence and player response. This cycle resembles scientific method, but it must operate at the speed of a competitive season.

Confidence affects performance because it influences attention, risk tolerance, and emotional control. Under pressure, players may narrow their focus, rush decisions, or avoid demanding actions. Psychological preparation helps players maintain useful attention and commit to appropriate decisions in unstable moments.

Consider a defender returning from a hamstring injury. Laboratory strength tests are useful, but they do not fully represent the demands of chasing an attacker, braking sharply, turning under pressure, and clearing the ball while fatigued. Return-to-play should therefore progress from controlled tests to football-specific exposure.

Research evidence should be read with attention to sample size, playing level, age group, sex, competitive context, and measurement technology. Findings from elite male professionals do not always transfer directly to youth football, amateur environments, or women's football.

Football injuries emerge from the interaction between tissue capacity, movement demands, fatigue, contact, history, and training load. No single variable explains risk. Prevention therefore requires a system that combines strength, neuromuscular control, appropriate workload progression, recovery, and clear communication between coaches and medical staff.

In practical terms, hamstring and groin injury mechanisms should be understood as part of the larger performance ecology of football. The scientific task is to describe the relevant mechanisms, while the coaching task is to transform that knowledge into training, selection, communication, and match decisions.

8.3 Knee, Ankle, and Contact Injuries

Decision-making in football is time pressured and information rich. Players must scan before receiving, anticipate movement, recognize patterns, and decide while opponents attempt to hide cues or create deception. Cognitive speed is therefore inseparable from technical speed.

Confidence affects performance because it influences attention, risk tolerance, and emotional control. Under pressure, players may narrow their focus, rush decisions, or avoid demanding actions. Psychological preparation helps players maintain useful attention and commit to appropriate decisions in unstable moments.

Team cohesion is not a vague emotional quality. It appears in communication, pressing commitment, recovery runs, role acceptance, and the willingness to act for collective benefit. A team with strong cohesion can execute demanding tactical plans more consistently.

Context should protect practitioners from false certainty. A metric that is useful for one team model may be misleading for another. A player who appears inefficient in one tactical structure may become highly valuable in a different role. Scientific interpretation must therefore remain conditional and specific.

Return-to-play decisions should not be based only on the absence of pain. A player may be pain-free yet not ready for the speed, contact, fatigue, and decision-making demands of competition. The final stage of rehabilitation must recreate football-specific stress in a controlled progression.

Consider an academy player who dominates because of early maturation. The scouting report should distinguish current physical advantage from adaptable skill, perception, learning behaviour, and resilience. Long-term development depends on identifying qualities that will still matter when physical differences narrow.

Thresholds for speed zones, workload alerts, and fatigue markers vary across studies and commercial systems. Practitioners should understand how a threshold is defined before comparing numbers across teams, leagues, or seasons.

Kicking is a whole-body action. The support foot, pelvis, trunk, hip, knee, and ankle contribute to ball speed, accuracy, and trajectory. Technical coaching that focuses only on the striking leg misses the chain of coordination that produces a powerful and controlled strike.

In practical terms, knee, ankle, and contact injuries should be understood as part of the larger performance ecology of football. The scientific task is to describe the relevant mechanisms, while the coaching task is to transform that knowledge into training, selection, communication, and match decisions.

8.4 Neuromuscular Warm-Up Programmes

Football injuries emerge from the interaction between tissue capacity, movement demands, fatigue, contact, history, and training load. No single variable explains risk. Prevention therefore requires a system that combines strength, neuromuscular control, appropriate workload progression, recovery, and clear communication between coaches and medical staff.

Hamstring injuries are common because football repeatedly exposes players to sprinting, high-speed running, and rapid changes in muscle length. Prevention strategies usually combine sprint exposure, eccentric strength, progressive loading, and careful monitoring after congested match periods.

Return-to-play decisions should not be based only on the absence of pain. A player may be pain-free yet not ready for the speed, contact, fatigue, and decision-making demands of competition. The final stage of rehabilitation must recreate football-specific stress in a controlled progression.

For applied work, the central question is not whether the variable can be measured, but whether it changes decisions. A coach, analyst, or sports scientist should ask how the information affects training design, match preparation, player development, or medical risk management. Measurement without decision relevance adds administrative burden rather than performance value.

Biomechanics explains how forces are produced, transferred, absorbed, and redirected during football actions. Running, kicking, heading, landing, and turning are not merely matters of effort; they are coordinated movement solutions. Efficient technique allows the player to produce speed or power while limiting unnecessary stress on joints and soft tissue.

Consider a wide player who repeatedly makes high-speed runs behind the defensive line. A physical report may describe sprint distance, but tactical analysis must ask whether the runs were timed well, whether teammates had passing access, whether the runs stretched the opponent, and whether the player could repeat them after fatigue. The scientific value appears when physical output is connected to tactical effect.

Research evidence should be read with attention to sample size, playing level, age group, sex, competitive context, and measurement technology. Findings from elite male professionals do not always transfer directly to youth football, amateur environments, or women's football.

Aerobic fitness supports football because it helps players recover between explosive efforts, maintain technical quality during fatigue, and tolerate demanding training blocks. However, football is not a simple endurance sport. The decisive moments frequently arise from brief accelerations, rapid changes of direction, high-speed runs into space, and repeated pressing actions.

In practical terms, neuromuscular warm-up programmes should be understood as part of the larger performance ecology of football. The scientific task is to describe the relevant mechanisms, while the coaching task is to transform that knowledge into training, selection, communication, and

match decisions.

8.5 Return-to-Play Decision-Making

Biomechanics explains how forces are produced, transferred, absorbed, and redirected during football actions. Running, kicking, heading, landing, and turning are not merely matters of effort; they are coordinated movement solutions. Efficient technique allows the player to produce speed or power while limiting unnecessary stress on joints and soft tissue.

Acceleration requires forward lean, effective ground contact, and rapid force production. Deceleration requires the opposite skill: the capacity to brake safely and reorient the body. In football, braking may be as important as sprinting because tactical success often depends on stopping, turning, and acting again before an opponent can respond.

Kicking is a whole-body action. The support foot, pelvis, trunk, hip, knee, and ankle contribute to ball speed, accuracy, and trajectory. Technical coaching that focuses only on the striking leg misses the chain of coordination that produces a powerful and controlled strike.

A practical football environment also requires shared language. Coaches may describe a situation tactically, medical staff may describe it through load or risk, analysts may describe it through spatial data, and players may describe it through feel. The interdisciplinary task is to translate these perspectives into a coherent plan.

Aerobic fitness supports football because it helps players recover between explosive efforts, maintain technical quality during fatigue, and tolerate demanding training blocks. However, football is not a simple endurance sport. The decisive moments frequently arise from brief accelerations, rapid changes of direction, high-speed runs into space, and repeated pressing actions.

Consider a central midfielder who completes many short passes. A raw passing percentage may look excellent, but the analyst must examine pressure, pass direction, receiving body shape, and whether the passes progressed play. Football data becomes meaningful when it differentiates safe circulation from actions that alter the opponent's defensive structure.

Thresholds for speed zones, workload alerts, and fatigue markers vary across studies and commercial systems. Practitioners should understand how a threshold is defined before comparing numbers across teams, leagues, or seasons.

Decision-making in football is time pressured and information rich. Players must scan before receiving, anticipate movement, recognize patterns, and decide while opponents attempt to hide cues or create deception. Cognitive speed is therefore inseparable from technical speed.

In practical terms, return-to-play decision-making should be understood as part of the larger performance ecology of football. The scientific task is to describe the relevant mechanisms, while the coaching task is to transform that knowledge into training, selection, communication, and match decisions.

8.6 Medical Teams and Communication

Football performance depends on repeated high-intensity actions embedded within long periods of lower-intensity movement. Players walk, jog, accelerate, decelerate, sprint, jump, collide, and recover, often within short time windows. The energy cost of the game is therefore not represented adequately by total distance alone; the pattern and timing of intense actions are equally important.

Aerobic fitness supports football because it helps players recover between explosive efforts, maintain technical quality during fatigue, and tolerate demanding training blocks. However, football is not a simple endurance sport. The decisive moments frequently arise from brief accelerations, rapid changes of direction, high-speed runs into space, and repeated pressing actions.

Physiological preparation must respect positional demands. Full-backs and wide midfielders may require repeated high-speed running along corridors of the pitch, centre-backs may face intense accelerations and aerial duels, and central midfielders often accumulate large volumes of medium-intensity movement while constantly adjusting position.

The strongest programmes are iterative. They observe performance, identify problems, design interventions, test them in training, and adjust them according to evidence and player response. This cycle resembles scientific method, but it must operate at the speed of a competitive season.

Team cohesion is not a vague emotional quality. It appears in communication, pressing commitment, recovery runs, role acceptance, and the willingness to act for collective benefit. A team with strong cohesion can execute demanding tactical plans more consistently.

Consider a defender returning from a hamstring injury. Laboratory strength tests are useful, but they do not fully represent the demands of chasing an attacker, braking sharply, turning under pressure, and clearing the ball while fatigued. Return-to-play should therefore progress from controlled tests to football-specific exposure.

Research evidence should be read with attention to sample size, playing level, age group, sex, competitive context, and measurement technology. Findings from elite male professionals do not always transfer directly to youth football, amateur environments, or women's football.

Return-to-play decisions should not be based only on the absence of pain. A player may be pain-free yet not ready for the speed, contact, fatigue, and decision-making demands of competition. The final stage of rehabilitation must recreate football-specific stress in a controlled progression.

In practical terms, medical teams and communication should be understood as part of the larger performance ecology of football. The scientific task is to describe the relevant mechanisms, while the coaching task is to transform that knowledge into training, selection, communication, and match decisions.

Chapter 9

Psychology, Perception, and Decision-Making

This chapter explores cognition, emotion, and social behaviour in football. Performance depends on scanning, anticipation, attention, confidence, leadership, and the capacity to act intelligently under pressure.

9.1 Scanning and Visual Search

Technical skill in football is perception-action skill. A pass is not only a motor pattern but a decision made in relation to pressure, timing, space, and teammate movement. The best players do not merely execute techniques well; they select the appropriate technique earlier and under more difficult conditions.

Motor learning research supports practice environments that include variability, representative pressure, and meaningful information. Repetition remains important, but repetition without context may produce skill that appears clean in drills and fragile in matches. Practice should help players recognize cues, adapt body shape, and execute under realistic constraints.

The first touch is a critical bridge between receiving and acting. A good first touch can eliminate pressure, open the next pass, prepare a shot, or change the rhythm of an attack. A poor touch can compress time, invite pressing, and reduce the tactical options available to the team.

The strongest programmes are iterative. They observe performance, identify problems, design interventions, test them in training, and adjust them according to evidence and player response. This cycle resembles scientific method, but it must operate at the speed of a competitive season.

Pressing is not simply running at the ball. Effective pressing requires triggers, cover shadows, distances between lines, timing from supporting players, and a clear plan for what happens if the press is broken. Poorly organized pressing can create large spaces behind the first line of pressure.

Consider an academy player who dominates because of early maturation. The scouting report should distinguish current physical advantage from adaptable skill, perception, learning behaviour, and resilience. Long-term development depends on identifying qualities that will still matter when physical differences narrow.

The best evidence-informed practice combines published research, local data, practitioner expertise, and player feedback. None of these sources is sufficient on its own.

Data becomes valuable only when it supports better questions. A dashboard full of numbers can distract a coaching staff if the metrics are disconnected from the team's model of play. Good analysis should clarify decisions, not replace football judgement.

In practical terms, scanning and visual search should be understood as part of the larger performance ecology of football. The scientific task is to describe the relevant mechanisms, while the coaching task is to transform that knowledge into training, selection, communication, and match decisions.

9.2 Anticipation and Pattern Recognition

Tactics organize individual actions into collective behaviour. Formations provide reference points, but they do not describe the full reality of play. A team may defend in one structure, build attacks in another, press in a third, and alter spacing according to the opponent's behaviour.

Pressing is not simply running at the ball. Effective pressing requires triggers, cover shadows, distances between lines, timing from supporting players, and a clear plan for what happens if the press is broken. Poorly organized pressing can create large spaces behind the first line of pressure.

Transitions are decisive because teams are temporarily disorganized when possession changes. The first few seconds after losing or winning the ball can determine whether a team creates a chance, prevents a counterattack, or loses control of the match.

Context should protect practitioners from false certainty. A metric that is useful for one team model may be misleading for another. A player who appears inefficient in one tactical structure may become highly valuable in a different role. Scientific interpretation must therefore remain conditional and specific.

Data becomes valuable only when it supports better questions. A dashboard full of numbers can distract a coaching staff if the metrics are disconnected from the team's model of play. Good analysis should clarify decisions, not replace football judgement.

Consider a wide player who repeatedly makes high-speed runs behind the defensive line. A physical report may describe sprint distance, but tactical analysis must ask whether the runs were timed well, whether teammates had passing access, whether the runs stretched the opponent, and whether the player could repeat them after fatigue. The scientific value appears when physical output is connected to tactical effect.

Many football studies are observational. They describe relationships but do not always prove causality. A variable may correlate with success because good teams create the conditions for it, not because the variable alone produces success.

Confidence affects performance because it influences attention, risk tolerance, and emotional control. Under pressure, players may narrow their focus, rush decisions, or avoid demanding actions. Psychological preparation helps players maintain useful attention and commit to appropriate decisions in unstable moments.

In practical terms, anticipation and pattern recognition should be understood as part of the larger performance ecology of football. The scientific task is to describe the relevant mechanisms, while the coaching task is to transform that knowledge into training, selection, communication, and match decisions.

9.3 Pressure, Confidence, and Emotional Control

Performance analysis converts match events into interpretable information. Event data records actions such as passes, shots, tackles, and carries, while tracking data captures player and ball positions over time. Together they allow analysts to study not only what happened, but also where, when, and under which spatial conditions it happened.

Expected goals models estimate the probability that a shot becomes a goal based on historical patterns and shot characteristics. The metric is useful because it separates chance quality from finishing outcome, but it must be interpreted carefully. Models differ in their inputs, and a single match can be strongly shaped by variance.

Data becomes valuable only when it supports better questions. A dashboard full of numbers can distract a coaching staff if the metrics are disconnected from the team's model of play. Good analysis should clarify decisions, not replace football judgement.

For applied work, the central question is not whether the variable can be measured, but whether it changes decisions. A coach, analyst, or sports scientist should ask how the information affects training design, match preparation, player development, or medical risk management. Measurement without decision relevance adds administrative burden rather than performance value.

Decision-making in football is time pressured and information rich. Players must scan before receiving, anticipate movement, recognize patterns, and decide while opponents attempt to hide cues or create deception. Cognitive speed is therefore inseparable from technical speed.

Consider a central midfielder who completes many short passes. A raw passing percentage may look excellent, but the analyst must examine pressure, pass direction, receiving body shape, and whether the passes progressed play. Football data becomes meaningful when it differentiates safe circulation from actions that alter the opponent's defensive structure.

The best evidence-informed practice combines published research, local data, practitioner expertise, and player feedback. None of these sources is sufficient on its own.

Technical skill in football is perception-action skill. A pass is not only a motor pattern but a decision made in relation to pressure, timing, space, and teammate movement. The best players do not merely execute techniques well; they select the appropriate technique earlier and under more difficult conditions.

In practical terms, pressure, confidence, and emotional control should be understood as part of the larger performance ecology of football. The scientific task is to describe the relevant mechanisms, while the coaching task is to transform that knowledge into training, selection, communication, and match decisions.

9.4 Team Cohesion and Leadership

Decision-making in football is time pressured and information rich. Players must scan before receiving, anticipate movement, recognize patterns, and decide while opponents attempt to hide cues or create deception. Cognitive speed is therefore inseparable from technical speed.

Confidence affects performance because it influences attention, risk tolerance, and emotional control. Under pressure, players may narrow their focus, rush decisions, or avoid demanding actions. Psychological preparation helps players maintain useful attention and commit to appropriate decisions in unstable moments.

Team cohesion is not a vague emotional quality. It appears in communication, pressing commitment, recovery runs, role acceptance, and the willingness to act for collective benefit. A team with strong cohesion can execute demanding tactical plans more consistently.

A practical football environment also requires shared language. Coaches may describe a situation tactically, medical staff may describe it through load or risk, analysts may describe it through spatial data, and players may describe it through feel. The interdisciplinary task is to translate these perspectives into a coherent plan.

Motor learning research supports practice environments that include variability, representative pressure, and meaningful information. Repetition remains important, but repetition without context may produce skill that appears clean in drills and fragile in matches. Practice should help players recognize cues, adapt body shape, and execute under realistic constraints.

Consider a defender returning from a hamstring injury. Laboratory strength tests are useful, but they do not fully represent the demands of chasing an attacker, braking sharply, turning under pressure, and clearing the ball while fatigued. Return-to-play should therefore progress from controlled tests to football-specific exposure.

Many football studies are observational. They describe relationships but do not always prove causality. A variable may correlate with success because good teams create the conditions for it, not because the variable alone produces success.

Transitions are decisive because teams are temporarily disorganized when possession changes. The first few seconds after losing or winning the ball can determine whether a team creates a chance, prevents a counterattack, or loses control of the match.

In practical terms, team cohesion and leadership should be understood as part of the larger performance ecology of football. The scientific task is to describe the relevant mechanisms, while the coaching task is to transform that knowledge into training, selection, communication, and match decisions.

9.5 Mental Fatigue and Tactical Errors

Technical skill in football is perception-action skill. A pass is not only a motor pattern but a decision made in relation to pressure, timing, space, and teammate movement. The best players do not merely execute techniques well; they select the appropriate technique earlier and under more difficult conditions.

Motor learning research supports practice environments that include variability, representative pressure, and meaningful information. Repetition remains important, but repetition without context may produce skill that appears clean in drills and fragile in matches. Practice should help players recognize cues, adapt body shape, and execute under realistic constraints.

The first touch is a critical bridge between receiving and acting. A good first touch can eliminate pressure, open the next pass, prepare a shot, or change the rhythm of an attack. A poor touch can compress time, invite pressing, and reduce the tactical options available to the team.

The strongest programmes are iterative. They observe performance, identify problems, design interventions, test them in training, and adjust them according to evidence and player response. This cycle resembles scientific method, but it must operate at the speed of a competitive season.

Transitions are decisive because teams are temporarily disorganized when possession changes. The first few seconds after losing or winning the ball can determine whether a team creates a chance, prevents a counterattack, or loses control of the match.

Consider an academy player who dominates because of early maturation. The scouting report should distinguish current physical advantage from adaptable skill, perception, learning behaviour, and resilience. Long-term development depends on identifying qualities that will still matter when physical differences narrow.

The best evidence-informed practice combines published research, local data, practitioner expertise, and player feedback. None of these sources is sufficient on its own.

Expected goals models estimate the probability that a shot becomes a goal based on historical patterns and shot characteristics. The metric is useful because it separates chance quality from finishing outcome, but it must be interpreted carefully. Models differ in their inputs, and a single match can be strongly shaped by variance.

In practical terms, mental fatigue and tactical errors should be understood as part of the larger performance ecology of football. The scientific task is to describe the relevant mechanisms, while the coaching task is to transform that knowledge into training, selection, communication, and match decisions.

9.6 Coaching Communication and Learning Climate

Tactics organize individual actions into collective behaviour. Formations provide reference points, but they do not describe the full reality of play. A team may defend in one structure, build attacks in another, press in a third, and alter spacing according to the opponent's behaviour.

Pressing is not simply running at the ball. Effective pressing requires triggers, cover shadows, distances between lines, timing from supporting players, and a clear plan for what happens if the press is broken. Poorly organized pressing can create large spaces behind the first line of pressure.

Transitions are decisive because teams are temporarily disorganized when possession changes. The first few seconds after losing or winning the ball can determine whether a team creates a chance, prevents a counterattack, or loses control of the match.

Context should protect practitioners from false certainty. A metric that is useful for one team model may be misleading for another. A player who appears inefficient in one tactical structure may become highly valuable in a different role. Scientific interpretation must therefore remain conditional and specific.

Performance analysis converts match events into interpretable information. Event data records actions such as passes, shots, tackles, and carries, while tracking data captures player and ball positions over time. Together they allow analysts to study not only what happened, but also where, when, and under which spatial conditions it happened.

Consider a wide player who repeatedly makes high-speed runs behind the defensive line. A physical report may describe sprint distance, but tactical analysis must ask whether the runs were timed well, whether teammates had passing access, whether the runs stretched the opponent, and whether the player could repeat them after fatigue. The scientific value appears when physical output is connected to tactical effect.

Many football studies are observational. They describe relationships but do not always prove causality. A variable may correlate with success because good teams create the conditions for it, not because the variable alone produces success.

Decision-making in football is time pressured and information rich. Players must scan before receiving, anticipate movement, recognize patterns, and decide while opponents attempt to hide cues or create deception. Cognitive speed is therefore inseparable from technical speed.

In practical terms, coaching communication and learning climate should be understood as part of the larger performance ecology of football. The scientific task is to describe the relevant mechanisms, while the coaching task is to transform that knowledge into training, selection, communication, and match decisions.

Chapter 10

Youth Development and Talent Identification

This chapter discusses youth development and talent identification. It emphasizes maturation, learning environments, relative age effects, and the long-term difference between early success and sustainable potential.

10.1 Growth, Maturation, and Individual Pathways

Technical skill in football is perception-action skill. A pass is not only a motor pattern but a decision made in relation to pressure, timing, space, and teammate movement. The best players do not merely execute techniques well; they select the appropriate technique earlier and under more difficult conditions.

Motor learning research supports practice environments that include variability, representative pressure, and meaningful information. Repetition remains important, but repetition without context may produce skill that appears clean in drills and fragile in matches. Practice should help players recognize cues, adapt body shape, and execute under realistic constraints.

The first touch is a critical bridge between receiving and acting. A good first touch can eliminate pressure, open the next pass, prepare a shot, or change the rhythm of an attack. A poor touch can compress time, invite pressing, and reduce the tactical options available to the team.

Context should protect practitioners from false certainty. A metric that is useful for one team model may be misleading for another. A player who appears inefficient in one tactical structure may become highly valuable in a different role. Scientific interpretation must therefore remain conditional and specific.

Physiological preparation must respect positional demands. Full-backs and wide midfielders may require repeated high-speed running along corridors of the pitch, centre-backs may face intense accelerations and aerial duels, and central midfielders often accumulate large volumes of medium-intensity movement while constantly adjusting position.

Consider a central midfielder who completes many short passes. A raw passing percentage may look excellent, but the analyst must examine pressure, pass direction, receiving body shape, and whether the passes progressed play. Football data becomes meaningful when it differentiates safe circulation from actions that alter the opponent's defensive structure.

Research evidence should be read with attention to sample size, playing level, age group, sex, competitive context, and measurement technology. Findings from elite male professionals do not always transfer directly to youth football, amateur environments, or women's football.

Decision-making in football is time pressured and information rich. Players must scan before receiving, anticipate movement, recognize patterns, and decide while opponents attempt to hide cues or create deception. Cognitive speed is therefore inseparable from technical speed.

In practical terms, growth, maturation, and individual pathways should be understood as part of the larger performance ecology of football. The scientific task is to describe the relevant mechanisms, while the coaching task is to transform that knowledge into training, selection, communication, and match decisions.

10.2 The Relative Age Effect

Football performance depends on repeated high-intensity actions embedded within long periods of lower-intensity movement. Players walk, jog, accelerate, decelerate, sprint, jump, collide, and recover, often within short time windows. The energy cost of the game is therefore not represented adequately by total distance alone; the pattern and timing of intense actions are equally important.

Aerobic fitness supports football because it helps players recover between explosive efforts, maintain technical quality during fatigue, and tolerate demanding training blocks. However, football is not a simple endurance sport. The decisive moments frequently arise from brief accelerations, rapid changes of direction, high-speed runs into space, and repeated pressing actions.

Physiological preparation must respect positional demands. Full-backs and wide midfielders may require repeated high-speed running along corridors of the pitch, centre-backs may face intense accelerations and aerial duels, and central midfielders often accumulate large volumes of medium-intensity movement while constantly adjusting position.

For applied work, the central question is not whether the variable can be measured, but whether it changes decisions. A coach, analyst, or sports scientist should ask how the information affects training design, match preparation, player development, or medical risk management. Measurement without decision relevance adds administrative burden rather than performance value.

Decision-making in football is time pressured and information rich. Players must scan before receiving, anticipate movement, recognize patterns, and decide while opponents attempt to hide cues or create deception. Cognitive speed is therefore inseparable from technical speed.

Consider a defender returning from a hamstring injury. Laboratory strength tests are useful, but they do not fully represent the demands of chasing an attacker, braking sharply, turning under pressure, and clearing the ball while fatigued. Return-to-play should therefore progress from controlled tests to football-specific exposure.

Thresholds for speed zones, workload alerts, and fatigue markers vary across studies and commercial systems. Practitioners should understand how a threshold is defined before comparing numbers across teams, leagues, or seasons.

Developing players need technical repetition, tactical understanding, physical literacy, and psychological safety. Excessive early specialization or adult-style pressure can damage long-term growth. The aim is not to produce the best twelve-year-old, but to support the emergence of capable senior footballers.

In practical terms, the relative age effect should be understood as part of the larger performance ecology of football. The scientific task is to describe the relevant mechanisms, while the coaching task is to transform that knowledge into training, selection, communication, and match decisions.

10.3 Technical Development in Youth Football

Decision-making in football is time pressured and information rich. Players must scan before receiving, anticipate movement, recognize patterns, and decide while opponents attempt to hide cues or create deception. Cognitive speed is therefore inseparable from technical speed.

Confidence affects performance because it influences attention, risk tolerance, and emotional control. Under pressure, players may narrow their focus, rush decisions, or avoid demanding actions. Psychological preparation helps players maintain useful attention and commit to appropriate decisions in unstable moments.

Team cohesion is not a vague emotional quality. It appears in communication, pressing commitment, recovery runs, role acceptance, and the willingness to act for collective benefit. A team with strong cohesion can execute demanding tactical plans more consistently.

A practical football environment also requires shared language. Coaches may describe a situation tactically, medical staff may describe it through load or risk, analysts may describe it through spatial data, and players may describe it through feel. The interdisciplinary task is to translate these perspectives into a coherent plan.

The relative age effect describes the over-representation of players born early in a selection year. These players may appear more talented because they are older and more mature than peers in the same age category. Academies should design evaluation processes that protect late developers and avoid premature exclusion.

Consider an academy player who dominates because of early maturation. The scouting report should distinguish current physical advantage from adaptable skill, perception, learning behaviour, and resilience. Long-term development depends on identifying qualities that will still matter when physical differences narrow.

Research evidence should be read with attention to sample size, playing level, age group, sex, competitive context, and measurement technology. Findings from elite male professionals do not always transfer directly to youth football, amateur environments, or women's football.

Motor learning research supports practice environments that include variability, representative pressure, and meaningful information. Repetition remains important, but repetition without context may produce skill that appears clean in drills and fragile in matches. Practice should help players recognize cues, adapt body shape, and execute under realistic constraints.

In practical terms, technical development in youth football should be understood as part of the larger performance ecology of football. The scientific task is to describe the relevant mechanisms, while the coaching task is to transform that knowledge into training, selection, communication, and match decisions.

10.4 Physical Preparation for Developing Players

Youth development must account for biological maturation. Two players of the same chronological age may differ substantially in height, strength, speed, coordination, and emotional readiness. Talent identification that ignores maturation may reward temporary physical advantage rather than long-term football potential.

The relative age effect describes the over-representation of players born early in a selection year. These players may appear more talented because they are older and more mature than peers in the same age category. Academies should design evaluation processes that protect late developers and avoid premature exclusion.

Developing players need technical repetition, tactical understanding, physical literacy, and psychological safety. Excessive early specialization or adult-style pressure can damage long-term growth. The aim is not to produce the best twelve-year-old, but to support the emergence of capable senior footballers.

The strongest programmes are iterative. They observe performance, identify problems, design interventions, test them in training, and adjust them according to evidence and player response. This cycle resembles scientific method, but it must operate at the speed of a competitive season.

The first touch is a critical bridge between receiving and acting. A good first touch can eliminate pressure, open the next pass, prepare a shot, or change the rhythm of an attack. A poor touch can compress time, invite pressing, and reduce the tactical options available to the team.

Consider a wide player who repeatedly makes high-speed runs behind the defensive line. A physical report may describe sprint distance, but tactical analysis must ask whether the runs were timed well, whether teammates had passing access, whether the runs stretched the opponent, and whether the player could repeat them after fatigue. The scientific value appears when physical output is connected to tactical effect.

Thresholds for speed zones, workload alerts, and fatigue markers vary across studies and commercial systems. Practitioners should understand how a threshold is defined before comparing numbers across teams, leagues, or seasons.

Football performance depends on repeated high-intensity actions embedded within long periods of lower-intensity movement. Players walk, jog, accelerate, decelerate, sprint, jump, collide, and recover, often within short time windows. The energy cost of the game is therefore not represented adequately by total distance alone; the pattern and timing of intense actions are equally important.

In practical terms, physical preparation for developing players should be understood as part of the larger performance ecology of football. The scientific task is to describe the relevant mechanisms, while the coaching task is to transform that knowledge into training, selection, communication, and match decisions.

10.5 Education, Welfare, and Long-Term Development

Technical skill in football is perception-action skill. A pass is not only a motor pattern but a decision made in relation to pressure, timing, space, and teammate movement. The best players do not merely execute techniques well; they select the appropriate technique earlier and under more difficult conditions.

Motor learning research supports practice environments that include variability, representative pressure, and meaningful information. Repetition remains important, but repetition without context may produce skill that appears clean in drills and fragile in matches. Practice should help players recognize cues, adapt body shape, and execute under realistic constraints.

The first touch is a critical bridge between receiving and acting. A good first touch can eliminate pressure, open the next pass, prepare a shot, or change the rhythm of an attack. A poor touch can compress time, invite pressing, and reduce the tactical options available to the team.

Context should protect practitioners from false certainty. A metric that is useful for one team model may be misleading for another. A player who appears inefficient in one tactical structure may become highly valuable in a different role. Scientific interpretation must therefore remain conditional and specific.

Football performance depends on repeated high-intensity actions embedded within long periods of lower-intensity movement. Players walk, jog, accelerate, decelerate, sprint, jump, collide, and recover, often within short time windows. The energy cost of the game is therefore not represented adequately by total distance alone; the pattern and timing of intense actions are equally important.

Consider a central midfielder who completes many short passes. A raw passing percentage may look excellent, but the analyst must examine pressure, pass direction, receiving body shape, and whether the passes progressed play. Football data becomes meaningful when it differentiates safe circulation from actions that alter the opponent's defensive structure.

Research evidence should be read with attention to sample size, playing level, age group, sex, competitive context, and measurement technology. Findings from elite male professionals do not always transfer directly to youth football, amateur environments, or women's football.

Team cohesion is not a vague emotional quality. It appears in communication, pressing commitment, recovery runs, role acceptance, and the willingness to act for collective benefit. A team with strong cohesion can execute demanding tactical plans more consistently.

In practical terms, education, welfare, and long-term development should be understood as part of the larger performance ecology of football. The scientific task is to describe the relevant mechanisms, while the coaching task is to transform that knowledge into training, selection, communication, and match decisions.

10.6 From Academy Promise to Senior Performance

Football performance depends on repeated high-intensity actions embedded within long periods of lower-intensity movement. Players walk, jog, accelerate, decelerate, sprint, jump, collide, and recover, often within short time windows. The energy cost of the game is therefore not represented adequately by total distance alone; the pattern and timing of intense actions are equally important.

Aerobic fitness supports football because it helps players recover between explosive efforts, maintain technical quality during fatigue, and tolerate demanding training blocks. However, football is not a simple endurance sport. The decisive moments frequently arise from brief accelerations, rapid changes of direction, high-speed runs into space, and repeated pressing actions.

Physiological preparation must respect positional demands. Full-backs and wide midfielders may require repeated high-speed running along corridors of the pitch, centre-backs may face intense accelerations and aerial duels, and central midfielders often accumulate large volumes of medium-intensity movement while constantly adjusting position.

For applied work, the central question is not whether the variable can be measured, but whether it changes decisions. A coach, analyst, or sports scientist should ask how the information affects training design, match preparation, player development, or medical risk management. Measurement without decision relevance adds administrative burden rather than performance value.

Confidence affects performance because it influences attention, risk tolerance, and emotional control. Under pressure, players may narrow their focus, rush decisions, or avoid demanding actions. Psychological preparation helps players maintain useful attention and commit to appropriate decisions in unstable moments.

Consider a defender returning from a hamstring injury. Laboratory strength tests are useful, but they do not fully represent the demands of chasing an attacker, braking sharply, turning under pressure, and clearing the ball while fatigued. Return-to-play should therefore progress from controlled tests to football-specific exposure.

Thresholds for speed zones, workload alerts, and fatigue markers vary across studies and commercial systems. Practitioners should understand how a threshold is defined before comparing numbers across teams, leagues, or seasons.

The relative age effect describes the over-representation of players born early in a selection year. These players may appear more talented because they are older and more mature than peers in the same age category. Academies should design evaluation processes that protect late developers and avoid premature exclusion.

In practical terms, from academy promise to senior performance should be understood as part of the larger performance ecology of football. The scientific task is to describe the relevant mechanisms, while the coaching task is to transform that knowledge into training, selection, communication, and match decisions.

Chapter 11

Women's Football and Performance Science

This chapter examines women's football as a rapidly developing performance environment. It highlights the need for specific research, appropriate support structures, and careful interpretation of data from the men's game.

11.1 The Growth of the Women's Game

Football injuries emerge from the interaction between tissue capacity, movement demands, fatigue, contact, history, and training load. No single variable explains risk. Prevention therefore requires a system that combines strength, neuromuscular control, appropriate workload progression, recovery, and clear communication between coaches and medical staff.

Hamstring injuries are common because football repeatedly exposes players to sprinting, high-speed running, and rapid changes in muscle length. Prevention strategies usually combine sprint exposure, eccentric strength, progressive loading, and careful monitoring after congested match periods.

Return-to-play decisions should not be based only on the absence of pain. A player may be pain-free yet not ready for the speed, contact, fatigue, and decision-making demands of competition. The final stage of rehabilitation must recreate football-specific stress in a controlled progression.

For applied work, the central question is not whether the variable can be measured, but whether it changes decisions. A coach, analyst, or sports scientist should ask how the

information affects training design, match preparation, player development, or medical risk management. Measurement without decision relevance adds administrative burden rather than performance value.

Football performance depends on repeated high-intensity actions embedded within long periods of lower-intensity movement. Players walk, jog, accelerate, decelerate, sprint, jump, collide, and recover, often within short time windows. The energy cost of the game is therefore not represented adequately by total distance alone; the pattern and timing of intense actions are equally important.

Consider an academy player who dominates because of early maturation. The scouting report should distinguish current physical advantage from adaptable skill, perception, learning behaviour, and resilience. Long-term development depends on identifying qualities that will still matter when physical differences narrow.

Many football studies are observational. They describe relationships but do not always prove causality. A variable may correlate with success because good teams create the conditions for it, not because the variable alone produces success.

Pressing is not simply running at the ball. Effective pressing requires triggers, cover shadows, distances between lines, timing from supporting players, and a clear plan for what happens if the press is broken. Poorly organized pressing can create large spaces behind the first line of pressure.

In practical terms, the growth of the women's game should be understood as part of the larger performance ecology of football. The scientific task is to describe the relevant mechanisms, while the coaching task is to transform that knowledge into training, selection, communication, and match decisions.

11.2 Sex-Specific Research Gaps

Football performance depends on repeated high-intensity actions embedded within long periods of lower-intensity movement. Players walk, jog, accelerate, decelerate, sprint, jump, collide, and recover, often within short time windows. The energy cost of the game is therefore not represented adequately by total distance alone; the pattern and timing of intense actions are equally important.

Aerobic fitness supports football because it helps players recover between explosive efforts, maintain technical quality during fatigue, and tolerate demanding training blocks. However, football is not a simple endurance sport. The decisive moments frequently arise from brief accelerations, rapid changes of direction, high-speed runs into space, and repeated pressing actions.

Physiological preparation must respect positional demands. Full-backs and wide midfielders may require repeated high-speed running along corridors of the pitch, centre-backs may face intense accelerations and aerial duels, and central midfielders often accumulate large volumes of medium-intensity movement while constantly adjusting position.

A practical football environment also requires shared language. Coaches may describe a situation tactically, medical staff may describe it through load or risk, analysts may describe it through spatial data, and players may describe it through feel. The interdisciplinary task is to translate these perspectives into a coherent plan.

Pressing is not simply running at the ball. Effective pressing requires triggers, cover shadows, distances between lines, timing from supporting players, and a clear plan for what happens if the press is broken. Poorly organized pressing can create large spaces behind the first line of pressure.

Consider a wide player who repeatedly makes high-speed runs behind the defensive line. A physical report may describe sprint distance, but tactical analysis must ask whether the runs were timed well, whether teammates had passing access, whether the runs stretched the opponent, and whether the player could repeat them after fatigue. The scientific value appears when physical output is connected to tactical effect.

The best evidence-informed practice combines published research, local data, practitioner expertise, and player feedback. None of these sources is sufficient on its own.

Women's football has grown rapidly, but research and support structures have often lagged behind the professionalization of the game. Performance science must not simply copy assumptions from men's football. It should study the actual demands, injury patterns, resources, and developmental pathways of female players.

In practical terms, sex-specific research gaps should be understood as part of the larger performance ecology of football. The scientific task is to describe the relevant mechanisms, while the coaching task is to transform that knowledge into training, selection, communication, and match decisions.

11.3 Training Load and Professional Support

Tactics organize individual actions into collective behaviour. Formations provide reference points, but they do not describe the full reality of play. A team may defend in one structure, build attacks in another, press in a third, and alter spacing according to the opponent's behaviour.

Pressing is not simply running at the ball. Effective pressing requires triggers, cover shadows, distances between lines, timing from supporting players, and a clear plan for what happens if the press is broken. Poorly organized pressing can create large spaces behind the first line of pressure.

Transitions are decisive because teams are temporarily disorganized when possession changes. The first few seconds after losing or winning the ball can determine whether a team creates a chance, prevents a counterattack, or loses control of the match.

The strongest programmes are iterative. They observe performance, identify problems, design interventions, test them in training, and adjust them according to evidence and player response. This cycle resembles scientific method, but it must operate at the speed of a competitive season.

The tactical and technical development of women's football is accelerating as professional environments improve. Better academies, coaching, analysis, nutrition, and sports medicine will likely continue to raise the speed, intensity, and sophistication of the game.

Consider a central midfielder who completes many short passes. A raw passing percentage may look excellent, but the analyst must examine pressure, pass direction, receiving body shape, and whether the passes progressed play. Football data becomes meaningful when it differentiates safe circulation from actions that alter the opponent's defensive structure.

Many football studies are observational. They describe relationships but do not always prove causality. A variable may correlate with success because good teams create the conditions for it, not because the variable alone produces success.

Return-to-play decisions should not be based only on the absence of pain. A player may be pain-free yet not ready for the speed, contact, fatigue, and decision-making demands of competition. The final stage of rehabilitation must recreate football-specific stress in a controlled progression.

In practical terms, training load and professional support should be understood as part of the larger performance ecology of football. The scientific task is to describe the relevant mechanisms, while the coaching task is to transform that knowledge into training, selection, communication, and match decisions.

11.4 Injury Patterns and Prevention Priorities

Women's football has grown rapidly, but research and support structures have often lagged behind the professionalization of the game. Performance science must not simply copy assumptions from men's football. It should study the actual demands, injury patterns, resources, and developmental pathways of female players.

Injury prevention in women's football requires attention to strength, neuromuscular control, landing mechanics, scheduling, menstrual-cycle-related considerations where appropriate, and access to qualified medical support. The priority is not to frame female athletes as fragile, but to build robust systems around their performance needs.

The tactical and technical development of women's football is accelerating as professional environments improve. Better academies, coaching, analysis, nutrition, and sports medicine will likely continue to raise the speed, intensity, and sophistication of the game.

Context should protect practitioners from false certainty. A metric that is useful for one team model may be misleading for another. A player who appears inefficient in one tactical structure may become highly valuable in a different role. Scientific interpretation must therefore remain conditional and specific.

Football injuries emerge from the interaction between tissue capacity, movement demands, fatigue, contact, history, and training load. No single variable explains risk. Prevention therefore requires a system that combines strength, neuromuscular control, appropriate workload progression, recovery, and clear communication between coaches and medical staff.

Consider a defender returning from a hamstring injury. Laboratory strength tests are useful, but they do not fully represent the demands of chasing an attacker, braking sharply, turning under pressure, and clearing the ball while fatigued. Return-to-play should therefore progress from controlled tests to football-specific exposure.

The best evidence-informed practice combines published research, local data, practitioner expertise, and player feedback. None of these sources is sufficient on its own.

Aerobic fitness supports football because it helps players recover between explosive efforts, maintain technical quality during fatigue, and tolerate demanding training blocks. However, football is not a simple endurance sport. The decisive moments frequently arise from brief accelerations, rapid changes of direction, high-speed runs into space, and repeated pressing actions.

In practical terms, injury patterns and prevention priorities should be understood as part of the larger performance ecology of football. The scientific task is to describe the relevant mechanisms, while the coaching task is to transform that knowledge into training, selection, communication, and match decisions.

11.5 Tactical and Technical Evolution

Football injuries emerge from the interaction between tissue capacity, movement demands, fatigue, contact, history, and training load. No single variable explains risk. Prevention therefore requires a system that combines strength, neuromuscular control, appropriate workload progression, recovery, and clear communication between coaches and medical staff.

Hamstring injuries are common because football repeatedly exposes players to sprinting, high-speed running, and rapid changes in muscle length. Prevention strategies usually combine sprint exposure, eccentric strength, progressive loading, and careful monitoring after congested match periods.

Return-to-play decisions should not be based only on the absence of pain. A player may be pain-free yet not ready for the speed, contact, fatigue, and decision-making demands of competition. The final stage of rehabilitation must recreate football-specific stress in a controlled progression.

For applied work, the central question is not whether the variable can be measured, but whether it changes decisions. A coach, analyst, or sports scientist should ask how the information affects training design, match preparation, player development, or medical risk management. Measurement without decision relevance adds administrative burden rather than performance value.

Aerobic fitness supports football because it helps players recover between explosive efforts, maintain technical quality during fatigue, and tolerate demanding training blocks. However, football is not a simple endurance sport. The decisive moments frequently arise from brief accelerations, rapid changes of direction, high-speed runs into space, and repeated pressing actions.

Consider an academy player who dominates because of early maturation. The scouting report should distinguish current physical advantage from adaptable skill, perception, learning behaviour, and resilience. Long-term development depends on identifying qualities that will still matter when physical differences narrow.

Many football studies are observational. They describe relationships but do not always prove causality. A variable may correlate with success because good teams create the conditions for it, not because the variable alone produces success.

Tactics organize individual actions into collective behaviour. Formations provide reference points, but they do not describe the full reality of play. A team may defend in one structure, build attacks in another, press in a third, and alter spacing according to the opponent's behaviour.

In practical terms, tactical and technical evolution should be understood as part of the larger performance ecology of football. The scientific task is to describe the relevant mechanisms, while the coaching task is to transform that knowledge into training, selection, communication, and match decisions.

11.6 Building Sustainable Performance Systems

Football performance depends on repeated high-intensity actions embedded within long periods of lower-intensity movement. Players walk, jog, accelerate, decelerate, sprint, jump, collide, and recover, often within short time windows. The energy cost of the game is therefore not represented adequately by total distance alone; the pattern and timing of intense actions are equally important.

Aerobic fitness supports football because it helps players recover between explosive efforts, maintain technical quality during fatigue, and tolerate demanding training blocks. However, football is not a simple endurance sport. The decisive moments frequently arise from brief accelerations, rapid changes of direction, high-speed runs into space, and repeated pressing actions.

Physiological preparation must respect positional demands. Full-backs and wide midfielders may require repeated high-speed running along corridors of the pitch, centre-backs may face intense accelerations and aerial duels, and central midfielders often accumulate large volumes of medium-intensity movement while constantly adjusting position.

A practical football environment also requires shared language. Coaches may describe a situation tactically, medical staff may describe it through load or risk, analysts may describe it through spatial data, and players may describe it through feel. The interdisciplinary task is to translate these perspectives into a coherent plan.

Transitions are decisive because teams are temporarily disorganized when possession changes. The first few seconds after losing or winning the ball can determine whether a team creates a chance, prevents a counterattack, or loses control of the match.

Consider a wide player who repeatedly makes high-speed runs behind the defensive line. A physical report may describe sprint distance, but tactical analysis must ask whether the runs were timed well, whether teammates had passing access, whether the runs stretched the opponent, and whether the player could repeat them after fatigue. The scientific value appears when physical output is connected to tactical effect.

The best evidence-informed practice combines published research, local data, practitioner expertise, and player feedback. None of these sources is sufficient on its own.

The tactical and technical development of women's football is accelerating as professional environments improve. Better academies, coaching, analysis, nutrition, and sports medicine will likely continue to raise the speed, intensity, and sophistication of the game.

In practical terms, building sustainable performance systems should be understood as part of the larger performance ecology of football. The scientific task is to describe the relevant mechanisms, while the coaching task is to transform that knowledge into training, selection, communication, and match decisions.

Chapter 12

Technology and the Future of Football Science

This chapter looks toward the future of football science. Technology can expand what clubs know about performance, but its value depends on interpretation, ethics, and integration into coaching practice.

12.1 Wearables, GPS, and Optical Tracking

Performance analysis converts match events into interpretable information. Event data records actions such as passes, shots, tackles, and carries, while tracking data captures player and ball positions over time. Together they allow analysts to study not only what happened, but also where, when, and under which spatial conditions it happened.

Expected goals models estimate the probability that a shot becomes a goal based on historical patterns and shot characteristics. The metric is useful because it separates chance quality from finishing outcome, but it must be interpreted carefully. Models differ in their inputs, and a single match can be strongly shaped by variance.

Data becomes valuable only when it supports better questions. A dashboard full of numbers can distract a coaching staff if the metrics are disconnected from the team's model of play. Good analysis should clarify decisions, not replace football judgement.

A practical football environment also requires shared language. Coaches may describe a situation tactically, medical staff may describe it through load or risk, analysts may describe it through spatial data, and players may describe it through feel. The interdisciplinary task is to translate these perspectives into a coherent plan.

Acceleration requires forward lean, effective ground contact, and rapid force production. Deceleration requires the opposite skill: the capacity to brake safely and reorient the body. In football, braking may be as important as sprinting because tactical success often depends on stopping, turning, and acting again before an opponent can respond.

Consider a central midfielder who completes many short passes. A raw passing percentage may look excellent, but the analyst must examine pressure, pass direction, receiving body shape, and whether the passes progressed play. Football data becomes meaningful when it differentiates safe circulation from actions that alter the opponent's defensive structure.

Thresholds for speed zones, workload alerts, and fatigue markers vary across studies and commercial systems. Practitioners should understand how a threshold is defined before comparing numbers across teams, leagues, or seasons.

The scientific study of football must balance reduction and integration. Reduction allows the researcher to isolate sprint speed, joint angle, oxygen uptake, or passing frequency. Integration asks how these variables function inside a match. Both approaches are necessary; the danger lies in treating an isolated variable as if it explains performance by itself.

In practical terms, wearables, gps, and optical tracking should be understood as part of the larger performance ecology of football. The scientific task is to describe the relevant mechanisms, while the coaching task is to transform that knowledge into training, selection, communication, and match decisions.

12.2 Artificial Intelligence in Scouting and Analysis

Biomechanics explains how forces are produced, transferred, absorbed, and redirected during football actions. Running, kicking, heading, landing, and turning are not merely matters of effort; they are coordinated movement solutions. Efficient technique allows the player to produce speed or power while limiting unnecessary stress on joints and soft tissue.

Acceleration requires forward lean, effective ground contact, and rapid force production. Deceleration requires the opposite skill: the capacity to brake safely and reorient the body. In football, braking may be as important as sprinting because tactical success often depends on stopping, turning, and acting again before an opponent can respond.

Kicking is a whole-body action. The support foot, pelvis, trunk, hip, knee, and ankle contribute to ball speed, accuracy, and trajectory. Technical coaching that focuses only on the striking leg misses the chain of coordination that produces a powerful and controlled strike.

The strongest programmes are iterative. They observe performance, identify problems, design interventions, test them in training, and adjust them according to evidence and player response. This cycle resembles scientific method, but it must operate at the speed of a competitive season.

The scientific study of football must balance reduction and integration. Reduction allows the researcher to isolate sprint speed, joint angle, oxygen uptake, or passing frequency. Integration asks how these variables function inside a match. Both approaches are necessary; the danger lies in treating an isolated variable as if it explains performance by itself.

Consider a defender returning from a hamstring injury. Laboratory strength tests are useful, but they do not fully represent the demands of chasing an attacker, braking sharply, turning under pressure, and clearing the ball while fatigued. Return-to-play should therefore progress from controlled tests to football-specific exposure.

Research evidence should be read with attention to sample size, playing level, age group, sex, competitive context, and measurement technology. Findings from elite male professionals do not always transfer directly to youth football, amateur environments, or women's football.

Artificial intelligence can support scouting, opposition analysis, injury-risk modelling, and individualized training design. However, algorithmic outputs are not neutral. They depend on data quality, modelling assumptions, and organizational incentives. Clubs must understand both the strengths and limitations of the systems they use.

In practical terms, artificial intelligence in scouting and analysis should be understood as part of the larger performance ecology of football. The scientific task is to describe the relevant mechanisms, while the coaching task is to transform that knowledge into training, selection, communication, and match decisions.

12.3 Automated Video and Real-Time Feedback

Football is a complex adaptive sport rather than a closed laboratory task. The same pass, sprint, tackle, or shot can have different value depending on the score, the opponent's shape, the phase of play, the time in the match, and the individual qualities of the players involved. Scientific analysis is therefore most useful when it respects the ecology of the game and connects measurement to tactical meaning.

A match is produced by interacting constraints: rules, pitch dimensions, weather, fatigue, the referee's interpretation, coaching strategy, crowd pressure, and the changing distribution of space. These constraints do not operate separately. They continuously reshape the options available to each player and determine whether a technical action is easy, risky, delayed, or impossible.

The scientific study of football must balance reduction and integration. Reduction allows the researcher to isolate sprint speed, joint angle, oxygen uptake, or passing frequency. Integration asks how these variables function inside a match. Both approaches are necessary; the danger lies in treating an isolated variable as if it explains performance by itself.

Context should protect practitioners from false certainty. A metric that is useful for one team model may be misleading for another. A player who appears inefficient in one tactical structure may become highly valuable in a different role. Scientific interpretation must therefore remain conditional and specific.

Technology will continue to change how football is observed, trained, and governed. Wearables, optical tracking, automated tagging, and artificial intelligence can reveal patterns that are difficult to see manually. Yet technology must remain subordinate to ethical practice, player welfare, and football intelligence.

Consider an academy player who dominates because of early maturation. The scouting report should distinguish current physical advantage from adaptable skill, perception, learning behaviour, and resilience. Long-term development depends on identifying qualities that will still matter when physical differences narrow.

Thresholds for speed zones, workload alerts, and fatigue markers vary across studies and commercial systems. Practitioners should understand how a threshold is defined before comparing numbers across teams, leagues, or seasons.

Performance analysis converts match events into interpretable information. Event data records actions such as passes, shots, tackles, and carries, while tracking data captures player and ball positions over time. Together they allow analysts to study not only what happened, but also where, when, and under which spatial conditions it happened.

In practical terms, automated video and real-time feedback should be understood as part of the larger performance ecology of football. The scientific task is to describe the relevant mechanisms, while the coaching task is to transform that knowledge into training, selection, communication, and match decisions.

12.4 Robotics, Simulation, and Biomechanical Modelling

Technology will continue to change how football is observed, trained, and governed. Wearables, optical tracking, automated tagging, and artificial intelligence can reveal patterns that are difficult to see manually. Yet technology must remain subordinate to ethical practice, player welfare, and football intelligence.

Artificial intelligence can support scouting, opposition analysis, injury-risk modelling, and individualized training design. However, algorithmic outputs are not neutral. They depend on data quality, modelling assumptions, and organizational incentives. Clubs must understand both the strengths and limitations of the systems they use.

The future coach will increasingly operate as a scientist-practitioner: someone able to interpret evidence, test ideas, collaborate with specialists, and translate complex information into clear training tasks. The best environments will combine rigorous analysis with human understanding.

For applied work, the central question is not whether the variable can be measured, but whether it changes decisions. A coach, analyst, or sports scientist should ask how the

information affects training design, match preparation, player development, or medical risk management. Measurement without decision relevance adds administrative burden rather than performance value.

Expected goals models estimate the probability that a shot becomes a goal based on historical patterns and shot characteristics. The metric is useful because it separates chance quality from finishing outcome, but it must be interpreted carefully. Models differ in their inputs, and a single match can be strongly shaped by variance.

Consider a wide player who repeatedly makes high-speed runs behind the defensive line. A physical report may describe sprint distance, but tactical analysis must ask whether the runs were timed well, whether teammates had passing access, whether the runs stretched the opponent, and whether the player could repeat them after fatigue. The scientific value appears when physical output is connected to tactical effect.

Research evidence should be read with attention to sample size, playing level, age group, sex, competitive context, and measurement technology. Findings from elite male professionals do not always transfer directly to youth football, amateur environments, or women's football.

Kicking is a whole-body action. The support foot, pelvis, trunk, hip, knee, and ankle contribute to ball speed, accuracy, and trajectory. Technical coaching that focuses only on the striking leg misses the chain of coordination that produces a powerful and controlled strike.

In practical terms, robotics, simulation, and biomechanical modelling should be understood as part of the larger performance ecology of football. The scientific task is to describe the relevant mechanisms, while the coaching task is to transform that knowledge into training, selection, communication, and match decisions.

12.5 Data Governance and Player Privacy

Performance analysis converts match events into interpretable information. Event data records actions such as passes, shots, tackles, and carries, while tracking data captures player and ball positions over time. Together they allow analysts to study not only what happened, but also where, when, and under which spatial conditions it happened.

Expected goals models estimate the probability that a shot becomes a goal based on historical patterns and shot characteristics. The metric is useful because it separates chance quality from finishing outcome, but it must be interpreted carefully. Models differ in their inputs, and a single match can be strongly shaped by variance.

Data becomes valuable only when it supports better questions. A dashboard full of numbers can distract a coaching staff if the metrics are disconnected from the team's model of play. Good analysis should clarify decisions, not replace football judgement.

A practical football environment also requires shared language. Coaches may describe a situation tactically, medical staff may describe it through load or risk, analysts may describe it through spatial data, and players may describe it through feel. The interdisciplinary task is to translate these perspectives into a coherent plan.

Kicking is a whole-body action. The support foot, pelvis, trunk, hip, knee, and ankle contribute to ball speed, accuracy, and trajectory. Technical coaching that focuses only on the striking leg misses the chain of coordination that produces a powerful and controlled strike.

Consider a central midfielder who completes many short passes. A raw passing percentage may look excellent, but the analyst must examine pressure, pass direction, receiving body shape, and whether the passes progressed play. Football data becomes meaningful when it differentiates safe circulation from actions that alter the opponent's defensive structure.

Thresholds for speed zones, workload alerts, and fatigue markers vary across studies and commercial systems. Practitioners should understand how a threshold is defined before comparing numbers across teams, leagues, or seasons.

A match is produced by interacting constraints: rules, pitch dimensions, weather, fatigue, the referee's interpretation, coaching strategy, crowd pressure, and the changing distribution of space. These constraints do not operate separately. They continuously reshape the options available to each player and determine whether a technical action is easy, risky, delayed, or impossible.

In practical terms, data governance and player privacy should be understood as part of the larger performance ecology of football. The scientific task is to describe the relevant mechanisms, while the coaching task is to transform that knowledge into training, selection, communication, and match decisions.

12.6 The Future Coach as Scientist-Practitioner

Biomechanics explains how forces are produced, transferred, absorbed, and redirected during football actions. Running, kicking, heading, landing, and turning are not merely matters of effort; they are coordinated movement solutions. Efficient technique allows the player to produce speed or power while limiting unnecessary stress on joints and soft tissue.

Acceleration requires forward lean, effective ground contact, and rapid force production. Deceleration requires the opposite skill: the capacity to brake safely and reorient the body. In football, braking may be as important as sprinting because tactical success often depends on stopping, turning, and acting again before an opponent can respond.

Kicking is a whole-body action. The support foot, pelvis, trunk, hip, knee, and ankle contribute to ball speed, accuracy, and trajectory. Technical coaching that focuses only on the striking leg misses the chain of coordination that produces a powerful and controlled strike.

The strongest programmes are iterative. They observe performance, identify problems, design interventions, test them in training, and adjust them according to evidence and player response. This cycle resembles scientific method, but it must operate at the speed of a competitive season.

Football is a complex adaptive sport rather than a closed laboratory task. The same pass, sprint, tackle, or shot can have different value depending on the score, the opponent's shape, the phase of play, the time in the match, and the individual qualities of the players involved. Scientific analysis is therefore most useful when it respects the ecology of the game and connects measurement to tactical meaning.

Consider a defender returning from a hamstring injury. Laboratory strength tests are useful, but they do not fully represent the demands of chasing an attacker, braking sharply, turning under pressure, and clearing the ball while fatigued. Return-to-play should therefore progress from controlled tests to football-specific exposure.

Research evidence should be read with attention to sample size, playing level, age group, sex, competitive context, and measurement technology. Findings from elite male professionals do not always transfer directly to youth football, amateur environments, or women's football.

Technology will continue to change how football is observed, trained, and governed. Wearables, optical tracking, automated tagging, and artificial intelligence can reveal patterns that are difficult to see manually. Yet technology must remain subordinate to ethical practice, player welfare, and football intelligence.

In practical terms, the future coach as scientist-practitioner should be understood as part of the larger performance ecology of football. The scientific task is to describe the relevant mechanisms, while the coaching task is to transform that knowledge into training, selection, communication, and match decisions.

Glossary of Key Terms

Acceleration. The increase in running speed over time, often decisive in the first metres of a duel, press, or attacking run.

Aerobic capacity. The ability to use oxygen to sustain activity and support recovery between high-intensity efforts.

Change of direction. A movement involving braking, reorientation, and re-acceleration, often under tactical and perceptual pressure.

Counter-pressing. The immediate attempt to regain possession after losing the ball, usually before the opponent can organize a counterattack.

Expected goals. A probabilistic estimate of chance quality based on shot characteristics and historical scoring patterns.

External load. The measurable physical work performed by a player, such as distance, accelerations, high-speed running, or sprint count.

Internal load. The physiological or perceptual response to work, such as heart rate, blood lactate, perceived exertion, or wellness status.

Microcycle. A short training-planning unit, usually one week, organized around match demands, recovery, tactical preparation, and adaptation.

Neuromuscular control. The coordination of muscles and joints to stabilize, move, brake, and absorb forces efficiently.

Periodization. The organization of training content over time to balance adaptation, readiness, and competition performance.

Pressing trigger. A cue that signals when a team should apply coordinated pressure, such as a poor touch, backward pass, or isolated receiver.

Representative practice. Training that preserves key information, pressure, spacing, and decision demands of the competitive environment.

Return to play. The staged process of restoring a player from injury to full competitive availability while managing recurrence risk.

Scanning. Active visual search before and during play to collect information about teammates, opponents, space, and pressure.

Tracking data. Spatial-temporal data showing the positions of players and/or the ball over time.

Selected References and Further Reading

Barengo, N. C., et al. (2014). The impact of the FIFA 11+ training program on injury prevention in football players: A systematic review. International Journal of Environmental Research and Public Health, 11(11), 11986-12000.

Bizzini, M., & Dvorak, J. (2015). FIFA 11+: An effective programme to prevent football injuries in various player groups worldwide-a narrative review. British Journal of Sports Medicine, 49(9), 577-579.

Bloomfield, J., Polman, R., & O'Donoghue, P. (2007). Physical demands of different positions in FA Premier League soccer. Journal of Sports Science & Medicine, 6(1), 63-70.

Boyne, M., et al. (2024). It's not all about power: A systematic review and meta-analysis comparing sex-based differences in kicking biomechanics in soccer. Sports Biomechanics, 23(12), 1-44.

Chmura, P., et al. (2018). Match outcome and running performance in different intensity ranges among elite soccer players. Biology of Sport, 35(2), 197-203.

Gualtieri, A., et al. (2023). High-speed running and sprinting in professional adult soccer: Current thresholds definition, match demands and training strategies. Sports Medicine - Open, 9, 29.

Harper, D. J., Carling, C., & Kiely, J. (2019). High-intensity acceleration and deceleration demands in elite team sports competitive match play: A systematic review and meta-analysis. Sports Medicine, 49, 1923-1947.

Kellis, E., & Katis, A. (2007). Biomechanical characteristics and determinants of instep soccer kick. Journal of Sports Science & Medicine, 6(2), 154-165.

Mead, J., O'Hare, A., & McMenemy, P. (2023). Expected goals in football: Improving model performance and demonstrating value. PLOS ONE, 18(4), e0282295.

Miguel, M., Oliveira, R., Loureiro, N., García-Rubio, J., & Ibáñez, S. J. (2021). Load measures in training/match monitoring in soccer: A systematic review. International Journal of Environmental Research and Public Health, 18(5), 2721.

Rhini, M., et al. (2024). The physical demands of the match according to playing positions in professional soccer players. Biology of Sport / PMC indexed article.

Tenga, A., Holme, I., Ronglan, L. T., & Bahr, R. (2010). Effect of playing tactics on goal scoring in Norwegian professional soccer. Journal of Sports Sciences, 28(3), 237-244.

Williams, A. M., & Ford, P. R. (2008). Expertise and expert performance in sport. International Review of Sport and Exercise Psychology, 1(1), 4-18.

Woods, C. T., et al. (2020). Talent identification and development in football: A systematic review. Sports Medicine - Open and related sport science literature.

Keywords

football science, soccer performance, sport physiology, biomechanics, match analysis, tactical analysis, high-speed running, expected goals, sports medicine, injury prevention, FIFA 11+, training methodology, youth development, women football, player monitoring, GPS tracking, artificial intelligence, coaching science, periodization, neuromuscular control

Publisher: Eliva Press Global Ltd

Email: info@elivapress.com

www.ingramcontent.com/pod-product-compliance
Lightning Source LLC
LaVergne TN
LVHW050822061126
851857LV00012B/932
* 9 7 8 9 9 9 9 3 4 9 1 8 5 *